The Best American Science and Nature Writing 2015

GUEST EDITORS OF
THE BEST AMERICAN SCIENCE
AND NATURE WRITING

2000 DAVID QUAMMEN
2001 EDWARD O. WILSON
2002 NATALIE ANGIER
2003 RICHARD DAWKINS
2004 STEVEN PINKER
2005 JONATHAN WEINER
2006 BRIAN GREENE
2007 RICHARD PRESTON
2008 JEROME GROOPMAN
2009 ELIZABETH KOLBERT
2010 FREEMAN DYSON
2011 MARY ROACH
2012 DAN ARIELY
2013 SIDDHARTHA MUKHERJEE
2014 DEBORAH BLUM
2015 REBECCA SKLOOT

The Best American Science and Nature Writing™ 2015

Edited and with an Introduction
by Rebecca Skloot

Tim Folger, Series Editor

A Mariner Original
HOUGHTON MIFFLIN HARCOURT
BOSTON • NEW YORK 2015

www.hmhco.com

ISSN 1530-1508
ISBN 978-0-544-28674-0

Printed in the United States of America
DOC 10 9 8 7 6 5 4 3 2 1

2014. Copyright © 2014 by Meera Subramanian. Reprinted by permission of Meera Subramanian.

"Curious" by Kim Todd. First published in *River Teeth*, Spring 2014. Copyright © 2014 by Kim Todd. Reprinted by permission of Kim Todd.

"The Aftershocks" by David Wolman. First published in *Matter*, August 24, 2014. Copyright © 2014 by David Wolman. Reprinted by permission of David Wolman.

"From Billions to None" by Barry Yeoman. First published in *Audubon*, May/June 2014. Copyright © 2014 by Barry Yeoman. Reprinted by permission of Barry Yeoman.

Contents

Foreword

WHEN ALBERT EINSTEIN was 16 years old and in his final year of high school, he performed an unusual experiment. He didn't use a laboratory, or any apparatus at all. Instead he conducted what may have been the first of his many *Gedankenexperimente*—thought experiments. He would continue to practice this imaginative yet rigorous sort of musing throughout his life, but in this particular case, the not-yet-iconic thinker wondered what a beam of light would look like if he was running alongside it at the same speed. Many years later, in his *Autobiographical Notes,* Einstein pointed to that first *Gedanken* moment as the origin of the ideas that have since transformed our understanding of the nature of space and time.

This year marks the 100th anniversary of the publication of Einstein's general theory of relativity (and the 15th anniversary of this anthology), so perhaps it's not a bad time to engage in some *Gedankenexperimente* of our own. Here's one: What if the world's political leaders met and engaged in the same caliber of discourse that scientists do, with the same spirit of collaborative problem solving? Granted, it's a proposition far less grounded in reality than Einstein's footrace with light, but let's set aside our incredulity for the moment.

First, our imaginary leaders might prioritize the real challenges facing the planet today, discuss possible solutions, and then—cue the derisive snorts—decide on a course of action and carry it out. Even climate change, the gravest threat facing us, would yield to this approach. We know the source of the problem—we're emitting too many planet-warming gases—and we're certainly smart enough to solve it, and at bargain-basement costs compared with the catastrophic price of inaction. Meera Subramanian's "The City and the Sea" is a remarkable testament to how much just one person can contribute to solving this problem; imagine what a whole roomful could do.

There is evidence that politicians have entertained—at least briefly—this same outlandish *Gedankenexperiment.* Some years ago Shimon Peres, the former president of Israel, toured the European Organization for Nuclear Research in Switzerland, where scientists from 113 countries conduct experiments. Inspired by his meeting with that international community, Peres told the assembled group of researchers (which included a Palestinian physicist) that perhaps the nation-state was obsolete and that the intellectual cooperation exemplified by the scientists at CERN could serve as a model for us all.

Conversely, what if the world's scientific community were to model itself after our political elite? Say scientists formed ideological camps that stymied the efforts of rivals, or denied, despite all the overwhelming supporting evidence, the truth of a theory. Or even worse, what if they waged war, perhaps beneath banners emblazoned with contested equations? Long live $E = mc^2$! Death to the $E = mc$ infidels! This much is certain: there would be no international collaborations and no inventions as remarkable as CERN's Large Hadron Collider, housed in a 17-mile-long circular tunnel, which allowed physicists to discover the Higgs boson, a particle so crucial to the architecture of the universe that without it none of us would exist. Nor would we have Dennis Overbye's delightful account of Peter Higgs, "A Pioneer as Elusive as His Particle," who predicted the existence of this all-important particle 51 years ago.

Sadly, it's not at all clear which *Gedanken* experiment is the more preposterous: that scientists would abandon reason or that politicians (and we who elect them) would embrace it. Perhaps even stranger than my thought experiment is David Wolman's "The Aftershocks," which tells the story of a benighted political vendetta against seven Italian scientists in the wake of an earthquake. Given the state of the world, then, it's no small miracle that something like CERN and its giant particle collider even exist. Costing more than $3 billion, the LHC was conceived and built not to generate profit but only to further our understanding of the laws that govern reality. It represents the pursuit of pure knowledge on the largest scale in the history of humanity.

That our civilization, for all its fractiousness, can still manage to build something like the LHC is a sign of enormous hope. Science is an inherently optimistic enterprise, the working assumption being that nature is comprehensible; mysteries can be solved; we can make things better. If we can design a machine like the LHC, which essentially recreates the conditions that existed in the first few instants of the universe, surely we can find a way past problems that we ourselves have brought about.

Late last year, while busy gathering stories to send to Rebecca Skloot, our brilliant guest editor, I received an e-mail from a reader who expressed some of these same feelings about the nature of science:

> I have been a fan of this series for years and used it quite a bit when I taught freshmen expository writing to science majors. It seems to me that content has become darker and less hopeful over this time. Of course, I understand that dark days may yet lie ahead and that science is not just a barometer for potential doom but also an agent for change. But for me science is something I have turned to when I have lost all faith in humanity. When I marvel at what telescopes have seen, the mysteries of quantum mechanics, and the philosophical quandaries raised by neuroscience, I get giddy. How bad can the human condition be if we can make these investigations? I suppose I would just like to see a bit more wonder—a bit more magic—in the content and less doom and gloom.

I think readers of this current volume will find in its pages stories of wonder as well as eloquent and necessary accounts of the world we are altering so profoundly. Within these pages you'll have close encounters not only with scientists but with crows, whales, and hyenas. One guarantee: there will be no shortage of food for *Gedanken*.

I try to read widely while searching for articles for this anthology, but without the help of readers, writers, and editors I would miss many good stories. So lend a hand and nominate your favorites for next year's anthology at http://timfolger.net/forums. I encourage writers to submit their own stories. The criteria for submissions and deadlines and the address to which entries should be sent can be found in the "news and announcements" forum on my website. Once again this year I'm offering an incentive to enlist readers to scour the nation in search of good science and nature writing. Send me an article that I haven't found, and if the article makes it into the anthology, I'll mail you a free copy of next year's edition. What do you think, Rebecca? Can I get you to sign those copies? I also encourage readers to use the forums to leave feedback about the collection and to discuss all things scientific. The best way for publications to guarantee that their articles are considered for inclusion in the anthology is to place me on their subscription list, using the address posted in the "news and announcements" forum.

I'd like to thank Rebecca Skloot for selecting such a wonderful collection of stories for this year's anthology. You won't find a better nonfiction book than her best-selling *The Immortal Life of Henrietta Lacks.* Once again this year I'm indebted to Naomi Gibbs and her colleagues at Houghton Mifflin Harcourt, who make this collection possible. And as always I'm most grateful of all to Anne Nolan, my beauteous wife. I hate to *Gedanke* where I would be without her.

TIM FOLGER

Introduction

A DECADE AGO, at the University of Pennsylvania vet school, I sat on a linoleum floor stroking my dog's head. She was in the 16th of what would become a 20-year life, and she'd just had a small tumor removed from her leg. As she fought to keep her eyes open through her post-anesthetic fog, a veterinarian walked into the room, surgical mask dangling from his chin. He pulled a pair of latex gloves from his hands with two loud snaps, and a woman's voice called out to him from behind a computer screen.

"How'd it go?" she asked.

"Great," he said. "Patient's up, swimming around."

Without breaking stride, the vet tossed his gloves in a trash can and walked toward an exit.

"Wait, what?" I said from the floor. "Your patient's swimming?"

He nodded.

"What's your patient?" I asked.

"Goldfish," he said, as if operating on a fish was something as ordinary as spaying a dog or cat. Then he reached for the door.

"Your patient is a goldfish?" I said. "What did you do to it?"

"Removed a tumor from its nose," he said as he opened the door and started to walk through.

"Wait!" I said, jumping from my dog's side and running toward him with a barrage of questions: *How do you anesthetize a fish? Who pays for this? What else do you do to fish? How common is this?*

As the vet answered my questions, I scribbled notes on the back of my dog's surgery receipt. (You anesthetize a fish using a tub of water mixed with liquid anesthetic, a submersible pump, and a plastic tube that pumps the water into the fish's mouth, over its gills, then back into the tub. Like a recirculating fountain. Fish vets do MRIs, CT scans, bone stabilization, bloodwork, you name it. If you can do it to a dog or cat, you can do it to a fish. People sometimes spend thousands of dollars treating fish they won at the fair or bought for less than $5. Because they love them.)

After getting the vet's contact information and a promise that I could observe his next fish surgery, I finally let him leave. He'd hardly passed through the door when I picked up my BlackBerry (it *was* a decade ago) and started typing an e-mail to my editor at the *New York Times Magazine.* Subject heading: "Whoa." A few hours later I had an assignment.

I knew I'd write about fish medicine the moment I heard the sentence "Patient's up, swimming around," because it was a clear example of something I call a "*what* moment." I can trace every story I've written back to one (often several) of these: a moment that grabs my attention and makes me stop and say, *Wait—what?*

Such as, *What? Did you just say your sergeant ordered you to volunteer for a research study on the effects of an experimental drug but didn't tell you what the study was for or what the risks might be?* (Indeed he did, and this wasn't uncommon or illegal.) Or *What? Did you just say you can identify a person's race using a DNA sample?* (Yep, and he'd built a business around doing so, even though the science didn't support his claims.)

My book, *The Immortal Life of Henrietta Lacks,* started with one moment in a biology class when I was 16: *What do you mean these cancer cells have been alive and growing in labs around the world since the 1950s even though the woman they came from died? And what do you mean her cells are one of the most important tools in medicine but no one knows anything about her except that she was black?*

I visit a lot of science-writing classes to talk with students, and I often tell them that one of the most important skills they can develop as young reporters is learning to recognize "*what* moments." They happen so often in life, and they're so easy to miss—you're busy thinking about a deadline or a class or when you have to pick your kid up from school—and it takes time to stop and say, *Wait—what?*, and then even more time to be truly present for the answer. But this is essential to science writing: following your curiosity, letting it guide you not just to stories but also through them, to wherever they need to go.

When I started researching Henrietta Lacks, I thought I was

writing a book about a woman and her amazing cells, but that changed when I talked to her daughter, Deborah. She told me she'd love it if someone wrote a book about her mother, so the world would know who she was and what her cells did for science. Then she paused, and her voice grew suddenly terrified. "But how do I know you're really a journalist?" she snapped. "How do I know you're not coming to steal my cells?"

"What?" I said. "Why would you think I'd be coming to steal your cells?" And with that, the questions driving my book grew from "Who was Henrietta Lacks and what did her cells do for science?" to include "And why would her daughter think I was pretending to be a writer in order to steal her cells?" It turned out that cells from Henrietta's children had been used in research without their knowledge, just as Henrietta's cells had been; that people had posed as journalists and lawyers to get all sorts of things from them—information, cells—and it had never worked out well for the Lacks family. That second "*what* moment" changed the story completely. In the end, it's not just the story of Henrietta and her cells, it's also (and perhaps most importantly) the story of the enduring impact those cells have had on her family.

"*What* moments" are all about wonder and what we can learn from it, and the stories in this year's *Best American Science and Nature Writing* are filled with them. In "The Big Kill," Elizabeth Kolbert talks to conservationists trying to stop the destruction of their native plant and bird life through the wholesale slaughter of invasive species. "Let's get rid of the lot," one character says. "Let's get rid of all the predators—all the damned mustelids, all the rats, all the possums." Wait . . . What? Mass killing as conservation? The result is an important story about the vast damage we humans cause to animals and the environment when we introduce invasive species and the extreme choices scientists face as they try to fix the problems we've caused.

In 1848, Phineas Gage survived an explosion on a railroad construction site that sent a metal spike though his skull. As the story goes, his personality changed completely after the accident; he lost his inhibitions and became aggressive, even lewd. By studying how the damage to Gage's frontal lobe changed his personality, scientists were finally able to learn what that part of the brain really does. Because of this, he's been trotted out as one of the most famous patients in neuroscience for over a century. When I first heard his story decades ago, I said the same thing most people say: Wait—what? He survived a giant metal spike through the skull? And a big hole in his brain? Thankfully, Sam Kean followed these questions. It's an amazing story, one that it turns out may be based in quite a bit of fiction.

In her essay "Curious," Kim Todd examines what she calls "the nature of the itch we call 'curiosity.'" This is the very core of the "*what* moment," those sudden glimpses of the unexpected that grab the imaginations of both writers and scientists, demanding investigation. "Curiosity can be as obsessive as hunger or lechery, swamping the senses," she writes. "Its subjects seem so frivolous: a baby giraffe, a dodo skeleton, the Surinam toad." But of course they're not frivolous, because through them we learn about ourselves and our world. "Intellectual curiosity sparks science, art, all kinds of innovation," she writes. "Here, in most of 21st-century North America, it is held in the highest esteem. For much of history, though, coveting the secrets of the world and mulling over mushrooms and vipers threatened to drag one from thoughts of God." As a preacher in the early 1600s warned, "Curiosity is the spiritual adultery of the soul. Curiosity is spiritual drunkenness." To which I say, *Sign me up*.

For Todd, that moment of curiosity's spark is the strange appearance of a Surinam toad. In Sheila Webster Boneham's "A Question of Corvids," it's a crow outside a hotel that seems to say "yeeees" when she asks if it's hungry. That moment leads her on a deeply researched journey through folklore and ornithology, from the U.S. to Ireland to the eastern Sierras, all culminating in a concise, touching natural history of the corvid family of birds. The spark for Rebecca Boyle's "The Health Effects of a World Without Darkness" was the moment she realized that she can't see the stars from where she lives because she's surrounded by too much artificial light. "After journeying millions of years, their light is swallowed by city glare and my porch lantern," she writes. "Those that make it through will still fail: not even bright Betelgeuse can outshine my iPhone. Yet I am an astronomy writer, a person who thinks about stars and planets all the time. What does my neglect of the night sky say about the rest of humanity?"

In "The Aftershocks," David Wolman follows the story of seven Italian scientists charged with involuntary manslaughter for failing to warn the public about an earthquake that killed 297 and injured thousands. "The claim," writes Wolman: "They had knowingly neglected their responsibility to inform the population about the risk at hand." The verdict: "For delivering 'inexact, incomplete, and contradictory information,' the scientists and engineers were found guilty of involuntary manslaughter. They each received a six-year prison sentence, pending appeal." What?! Scientists sentenced for not conveying earthquake risk? For conveying "inexact, incomplete, and contradictory" information? Good science is often all about the seemingly inexact process of putting forth theories, testing them, coming up with incomplete or contradictory data, revising your theories, then doing it all again as you whittle your ideas, hoping someday they'll become proven theories. And if scientists are being prosecuted over "inexact, incomplete, and contradictory" information, then watch out, science writers: Red wine is good for you! It's bad for you! Meat will kill you! Meat will make you live longer!

At its core, like several other stories in this collection, "The Aftershocks" is about the importance of clear and accurate science communication, the many points at which that communication can fail as it travels from scientists through the media to the public, and what's at stake when it goes wrong. It's also a sobering reminder of how little most people understand about the scientific process and the concepts of risk and probability.

This is a book filled with questions. What happens when your child is diagnosed with disease no one has ever heard of? Or when you try to unlock—and perhaps even change—traumatic memories? Of course good science and nature writing doesn't just ask *What?* It also asks things like *Why?* and *How?* and *At what cost?* In his story "Waiting for Light," Jake Abrahamson didn't just write about the fact that some villages in India still live without light, he also asked what impact that has on them, the ways in which they might get light, and what that might cost, financially, culturally, and environmentally. His story and Rebecca Boyle's together illustrate another important job of science writing: highlighting areas of science, technology, and nature that many take for granted while others have no access to it, and asking important questions about the dangers of either extreme: What does it mean, for humans and their environment, to live without access to light? Or to live with relentless inescapable light?

In "Desegregating Wilderness," Jourdan Imani Keith makes the essential connection between the Civil Rights Act and the Wilderness Act—two landmark laws that celebrated their 50th anniversaries in 2014—to explore the important questions of why access to nature is so often segregated along lines of color and class, what problems that causes, and how we can fix it. We have, she says, "a segregated wilderness, one in which the wild is hardest to reach for the people who, for historical reasons, still have fewer of the financial assets required to get there." And in "At Risk," Keith takes a group of urban teens to build trails in the wild, weaving a beautiful essay about "at-risk" youth who are as deserving of protection and access to wilderness as the at-risk salmon in the rivers she helps them explore.

People often think of science and writing as vastly different endeavors, but they're very much the same. They're both driven by curiosity, by noticing small moments—a single unexpected piece of data in an experiment, a sentence someone says in passing, a tiny crack in a rock face—and taking the time to see where those moments might lead, what larger stories they might uncover that can teach us about everything from the tiniest organism to the entire solar system. This is one thing all stories in this collection have in common: they're written by and about people who take the time, and often a substantial amount of risk, to follow curiosity wherever it might lead, so we can all learn from it.

Sometimes those risks mean months or years devoted to research without knowing where it might go or whether it will someday get published, relying only on personal credit cards and a belief that the story or data you're following is important. Sometimes it means tackling controversial topics for which there are no easy answers, like finding a balance point between free enterprise, environmental safety, and public health.

Sometimes the risks are emotional. In "No Risky Chances," Atul Gawande, a physician, asks one of the hardest questions of all: What does it mean to have a good death, and how can he help patients accomplish such a thing? He realized that rather than rattling off treatment options and outcome probabilities to a patient facing terminal ovarian cancer, as he'd been trained to do, he should ask questions like "What were her biggest fears and concerns? What goals were most important to her?" What was she willing to endure now "for the possibility of more time later"? For

Gawande and his patients, these questions aren't just about good medical care, they're about the importance of story: "Life is meaningful because it is a story," he writes. "No one ever really has control; physics and biology and accident ultimately have their way in our lives. But . . . we have room to act and shape our stories —although as we get older, we do so within narrower and narrower confines."

And sometimes science and nature writers risk their lives to follow important stories. In "Digging Through the World's Oldest Graveyard," Amy Maxmen and the scientists she writes about search for the origins of humanity amid warring tribes in Ethiopia, where paleontologists travel with "two hammers, two shovels, four rifles."

Sheri Fink, a physician and reporter, immersed herself in a Liberian Ebola clinic, "a place both ordinary and otherworldly," to show us the "the rhythms of a single day" in an Ebola outbreak. One patient said to her, "They told me I should be very mindful of others. No touching." But, she writes, "His bed, like the others in the unit, was in an 8-by-10-foot space separated from others by wood-framed walls of tarp, and he shared a latrine with other patients." He cried and told her, "It's too pathetic. I think the world needs to come." And through Fink's incredible eye for detail, and her willingness to go where few others would, she allowed the world to see precisely what he meant.

I desperately wish that every writer in this collection who took risks to tell important stories survived the year. Matthew Power was a fearless and talented young journalist. He reported on everything from natural disasters to war zones; he followed what *Men's Journal* called "one man's absurd quest to become the first person to walk the entire length of the Amazon River—floods, electric eels, and machete-wielding natives be damned." He went into Afghanistan to report on the Taliban's destruction of Buddha statues. As his former *Harper's* editor, Roger Hodge, told the *New York Times,* "He was always searching for the human truth beneath the sorry facts. He wanted to live it—live what these people were living." And he did just that, much to the world's benefit.

For his story included in this collection, "Blood in the Sand," this meant traveling to Costa Rica, into the center of a heartbreaking and deadly battle between turtle conservationists and poachers. Two months after this story ran in *Outside Magazine,* Matthew

Power collapsed and died from heatstroke while reporting a story about an explorer walking the length of the Nile. News of Power's death filled my Facebook feed as so many mutual friends mourned his loss. We also mourned the incredible stories we lost with him, those "*what* moments" he would have noticed, stories that would have grabbed his vast curiosity, stories that perhaps only he would have risked following.

Writers aren't the only ones taking risks for these stories. I think I hardly breathed while reading Burkhard Bilger's "In Deep," which tells the story of a team working to map the deepest caves in the world: "On any given day, the cave might be home to a particle physicist from Berkeley, a molecular biologist from Russia, a spacecraft engineer from Washington, D.C., a rancher from Mexico, a geologist from Sweden, a tree surgeon from Colorado, a mathematician from Slovenia, a theater director from Poland, and a cave guide from Canada who lived in a Jeep and spent two hundred days a year underground," he wrote. "They were a paradoxical breed: restlessly active yet fond of tight places, highly analytical yet indifferent to risk . . . As far as I could tell, only two things truly connected them: a love of the unknown and a tolerance for pain." Bilger's vivid writing transports readers deep underground and brings those risks, and the characters who take them, to life.

Like those cave explorers, Cindy Lee Van Dover, the scientist Brooke Jarvis writes about in "The Deepest Dig," takes incredible risks for her research. She sinks for more than an hour in a submersible to get to the bottom of the ocean. "The view from its portholes moves through a spectrum of glowing greens and blues, eventually fading to pure black," Jarvis writes. "The only break from the darkness comes when the sub drops through clusters of bioluminescence that look like stars in the Milky Way. They're the only way for Van Dover to tell, in the complete darkness and absence of acceleration, that she's sinking at all." She lands in "a strange land of underwater volcanoes and mountain ranges, of vast plains and smoking basalt spires," where she's found, among other things, "concentrations of metals — gold, copper, nickel, and silver, as well as more esoteric minerals used in electronics — that make the richest mines on dry land look meager." And as Jarvis writes, "Where there's metal, there are miners, even at the bottom of the world." It's a story of fascinating science and the risks required to uncover it, but it's also about the risks—and potential benefits—of the brand-new industry of deep-sea mining.

I've been a fan of this series since its first edition, which my father bought me as a present in 2000. I was in graduate school, just one year into the decade it would take me to write *The Immortal Life*, and I haven't missed a single edition since. I keep my *Best American Science and Nature Writing* collection on a special shelf near my desk, and over the years I've turned to it time and time again for inspiration, entertainment, and education—my own, and that of my students. So I was giddy with excitement when Houghton Mifflin Harcourt asked me to be the series editor this year. Giddy, but daunted.

I'm a science person; I think in terms of data collection and sample size. When asked to rule on the best science writing of 2014, I set out to find and read every such story published, gathering as broad a data set as possible before drawing conclusions. And here's what I found: while reading a year's worth of writing about science and nature—with stories of drought, widespread disease, environmental destruction, overfishing, poaching—it's easy to despair about the future of our planet and all species on it. But I did come away feeling hope for the future of one species: the science and nature writer.

Though the health of the world they're reporting on is in a fragile state, the science and nature writers of 2014 left me feeling hopeful about human ingenuity, the wonder of science, and our ability to harness it to solve big problems (of our own creation and otherwise). The day after finalizing the selection of stories for this collection, I flew from Chicago to San Francisco, and along the way I saw at first hand the incredible drought we're facing in this country. As I flew over drying-up reservoirs, lakes, and aqueducts, I thought of Rowan Jacobsen's "Down by the River" and Meera Subramanian's "The City and the Sea," both stories of communities finding meaningful recovery from water-related disasters, through individual creativity, cooperation between groups too often at odds, and a spirit of working with the forces of nature rather than against them.

I'm relieved by the number of outlets (some old, many new) publishing strong science and nature writing. Also by the number of talented writers entering the field, particularly women and minorities, groups that have been underrepresented in all areas of science for too long.

After reading hundreds of science and nature stories, I eventually realized that the task I'd set out for myself—to find and read every single one published in 2014—was impossible. Many are online; many aren't. The amazing Tim Folger, in addition to writing tremendous science and nature pieces of his own, gathered stories throughout the year and narrowed them to a group of finalists. Despite his help and my own deep searching, I'm sure I missed some. But this is good news. It gives me great hope to think that I found so many wonderful examples of science and nature writing—far more than I could include here—and that there are surely others out there I didn't uncover. It also makes me curious to find them.

REBECCA SKLOOT

The Best American Science and Nature Writing 2015

JAKE ABRAHAMSON

Waiting for Light

FROM *Sierra*

THE SKY ABOVE northeast India looked like mango skin. It was late afternoon in May, and across a constellation of villages, deliverymen worked to unload their solar-charged lanterns from trucks and bicycles before nightfall. They leaned into dung-and-straw huts, calling, "Lantern, lantern." They passed the devices to women in colorful saris, to bucktoothed kids, to men in sweat-stained undershirts, lingering while the customer made sure the lantern worked, head angled and skeptical.

As the sky dimmed, the lanterns were hung from the ceiling of every shop, precise white spheres in the darkness. A mustard grinder ran his seeds through his diesel-powered machine. A bangle maker massaged heat into metal until the smell of malleability crept out. Children did homework. Women cooked dal. All were warmed by the day's trapped energy diffusing from the mud bricks. The next morning the deliverymen would retrieve the lanterns and whisk them back to solar-powered plants to be recharged during the day.

The rechargeable LED lanterns came in 2012, by way of Delhi-based solar energy concern Omnigrid Micropower Company (OMC), which leases them to about 36,000 customers across the state of Uttar Pradesh for $2.50 a month. When the company first set up shop, its primary focus was using solar plants to power telecom towers, which once depended on diesel generators. But that changed quickly. "As we looked at it closer, we realized the power demand would be larger for the community than for telecom," said Pär Almqvist, OMC's chief marketing officer. In addition to the lanterns, the company offers battery boxes that power fans and cell phones, and it will soon become the first Internet service provider in the area. But right now OMC's mission is all about light: put photons where there are none.

The market for OMC and its two dozen peer companies is huge. Roughly 400 million of India's 1.2 billion inhabitants lack access to electricity—more than the combined populations of the United States and Canada. Indians without electricity typically use kerosene wick lights, which cause eye and respiratory disease, start fires, rely on sporadically available fuel, and provide about the same amount of illumination as a birthday candle.

That May night I rode out of a village called Jangaon and into the hills. As Jangaon receded, its 800 lanterns blended in with the stars, and I saw in our headlights that the road was disappearing too. It was just whorls in sand now. I caught a passing glimpse of the pearly, vacant eyes of a buffalo. Then the brick wall of a tiny village appeared. This was Aat, an unlighted hamlet hidden amid mango orchards.

"This is the darkest place I have seen in my life," proclaimed the man sitting beside me in the back seat. He was Ritu Raj Verma, OMC's boss in the field, head of rollout, a person with an almost religious need to distribute light.

The car stopped, and Verma told the driver to cut the headlights. The prominent dichotomy in Verma's life was of light and dark, and he was determined that I see which prevailed here. "You see? Full dark," he said. "I will make this village glow like a sun. I will bring them the light." I couldn't see his face, but I saw his broad head, with a dollop of hair on its crown, silhouetted by the stars and moon out the window behind him.

Over the wall lived the family of Ramswarup Verma (no relation to Ritu Raj), a farmer of 40 with the grooved, leathery face of someone much older. He wakes and sleeps by the rise and fall of the sun, which isn't as romantic as it sounds. When night arrives and Aat gets dark, the Verma family goes to bed for lack of anything else to do.

Mr. Ramswarup, as he preferred to be called, didn't expect the electrical grid to reach Aat in his lifetime. "There are many larger villages. We only have four to five houses," he told me the next night, when I visited him with a translator. His biggest complaint about the darkness was that he couldn't visit the water pump at night. Without a light, he couldn't see the snakes.

"With a lantern I will go to the market, my children will study, I will get water from the ground," he said. "If I get the light, it will change my life."

We were on a dirt patio of sorts for his goats and buffalo, inside the brick wall but outside his house. We sipped chai. And just as I was viewing Aat through the prism of America, he was doing the reverse.

"How is there electricity twenty-four hours in the United States?" he asked me.

"Are there mosquitoes and flies in America?"

"Do you have mangoes?"

"Is there a dowry system?"

My translator, Manoj, asked if it was true that France is the only country without mosquitoes. I didn't know the answer.

Then the two of them were talking in Hindi, and Manoj was making a fist with his left hand and shining a flashlight on it with his right hand, slowly rotating the fist. "He is asking how when it is night in India, it is day in the United States."

I pointed to the night sky and said, "Did you know that a human has walked on the moon?"

Mr. Ramswarup shook his head, his expression unchanging. He didn't seem to care.

"Do you believe me?" I asked.

"Yes."

"What are the moon and stars to you?"

"They are the gods."

We drove away over the bumpy dirt road, which is impassable by OMC's bicycles and delivery trucks. The company is creating a program to deliver solar-charged batteries by motorbike. Later this year Mr. Ramswarup will lease a lantern that he can keep in his house and repower daily with a newly delivered battery.

For Indians who live in darkness, electricity has long been a mirage resolving just beyond the gridded horizon. In villages throughout the country, utility poles lean and split. Some hold wires, put up by the government decades ago and then abandoned; others were erected by well-to-do individuals who hoped the sight of grid pillars would compel a state body to string wires across them. The poles are a reminder of the dying notion that the only thing separating people from power is a few dozen miles of copper wire.

In fact, one obstacle to full electrification in India might be the institutional faith in the grid itself (or grids—there are four), which fails the consumers who *are* connected to it. India's grid routinely suffers 30 to 40 percent power loss between the source and end user, mostly from pilferage and technical inefficiencies, and the grid runs largely on coal.

The most promising avenue for powering up India's rural poor could lie with private companies like OMC, offering solar-powered light via lanterns, household panels, or tiny, self-contained grids. There are 15 to 20 such companies, plus dozens more hyper-local providers, something like the neighborhood barbershops of solar installment. As easily as OMC distributes a truckload of lanterns, another start-up can wire a few dozen buildings to a mini solar plant (called a minigrid) or install solar panels on a single rooftop. The different delivery methods have their own pros and cons, but one of off-grid solar's greatest advantages is the ease with which companies can tailor their offerings. In a likely scenario, companies would offer combinations of prepackaged lighting, minigrids, and rooftop solar panels.

A successful private-sector industry wouldn't only power up new consumers using renewables; it would also severely undermine any chance that these people would use dirty power down the road.

"We'll be building an entirely different system from the bottom up," said Justin Guay, associate director of the Sierra Club's International Climate Program.

When all is said and done, one of the coal industry's favorite narratives—that its ability to provide cheap power to people in need outweighs its nefarious effects—would be cut down by a counterargument that's 400 million strong.

But the industry needs help from India's government. Like start-ups in San Francisco and New York, many off-grid companies in India go belly-up before they can get on their feet. "A lot of the companies understand technology, but they don't understand how to maintain a relationship with a rural customer," said Sandhya Hegde of investment firm Khosla Impact, which recently put $2 million into BBOXX, a company similar to OMC that works in sub-Saharan Africa, where off-grid solar is booming.

The ones that last need continued financing. While international investors have taken notice of the industry's potential, domestic banks are hesitant to fund companies whose customers include the poorest people on earth. And those banks need to feel comfortable buying in if the industry is going to grow.

"The local banks are looking for 100 percent collateral, really high rates of interest, and short loan terms. That's a huge issue," said Alex Doukas, a research analyst at World Resources Institute. "If you really want these enterprises to scale up, you need to have financial institutions on the ground that understand the business models."

India's government is in a position to help, and with Prime Minister Narendra Modi's recent pledge to bring solar energy to all Indians who need it by 2019, Guay and Doukas hope he rides the cresting wave of off-grid solar by facilitating low-interest loans for both solar providers and their customers. Bangladesh did something similar in 2003, and off-grid has swept the countryside, with 80,000 rooftop systems now being installed a month—a rate that's rising.

A recent report coauthored by Guay valued the worldwide market for off-grid at $12 billion annually. That's for 1.3 billion potential customers who currently live without electricity, a majority of whom are in India and sub-Saharan Africa. After energy will come Internet access, fans, electric bicycles, and refrigerators.

Then again, the promise of energy for all is a tiresome one in India, where the government has been notoriously bumbling and corrupt. In the 1990s and 2000s, India made an attempt at off-grid solar. NGOs and government entities that no longer exist installed it on rooftops at a highly subsidized rate, and the broken remnants, too expensive for consumers to repair, still stick out of rooftops in Jangaon. But there were no OMCs then. A few years from now, there might be 80,000 new Indian households extending their lives beyond sunset per month. Or there won't. In either case, the people without power will be waiting for someone to make a decision that could change their lives in an instant.

The lanterns arrived in 15-year-old Bhawana Singh's village, just a few miles from Jangaon, in December 2012. They came in the typical way. After sundown one evening, without prior notice, a cart full of them rolled up to a prominent location in the village and was lit up. OMC's Ritu Raj Verma calls this a road show.

"We are illuminating forty, fifty lanterns on top of the vehicle," Verma said. "It looks like the sun rises there. Slowly, slowly, the villagers are attracted toward the light. We give them our lantern in their hand so they can enjoy this light for a fraction of a second."

A year and a half later, Bhawana Singh told me that she uses her family's OMC lantern for studying and crafting. Outside her thatched hut, she leaned in from the dark to catch the lantern's sphere of light and unrolled a colorful strip of bunting that she'd stitched together from discarded yarn and cement packaging. "I made this from the waste," she said.

"Will you sell that in the market?" I asked.

She went to a doorway and pretended to hang the bunting above it. "It is not for selling. It is for the house."

"Beautiful," said Verma.

"What did you think the first time you saw this lantern?" I asked.

"I very much enjoyed it. Now I can study. In the storms, it will not be off."

Before the lanterns arrived, she used a kerosene candle.

"Do you prefer the lantern?" I asked her.

A man sitting nearby jerked to life and pointed to his eyes, indicating the way kerosene fumes made them tear up. Bhawana nodded. Then Verma got into a conversation with one of the Singh men.

I asked him what they were talking about, and he said, "This Malaysian aircraft has been lost. I am telling them that when I would like to access information about that plane, I'm accessing my Internet. I'm not getting a newspaper from Malaysia."

"What else?"

"He is requesting that I create a library for the students over here. But the Internet is the biggest library. It's the electronic library."

"Do you hope that one day you will have electricity?" I asked Bhawana.

"Every year we think electricity will come," she said. "We think the light will come. Only the pillars are there, but there is no electricity in the wires."

"How long have the pillars been there?"

"Four years. Every year since they came we hope they will be turned on."

"It is the hope of a human being," Verma added. "Something

new is there, and sometime, someday, it will work." He was getting protective. He had a paternal thing with Bhawana, had kind of taken her under his wing.

Looking around the sky, I found the pillars that Bhawana was talking about—towering, solitary poles. Their silhouettes leaned like giant cacti against the moon.

Bhawana's younger sister Sonal sat down beside her. She held a painting of a rose that she had made. "What do you love in the world?" I asked.

"We love to study. We love to do embroidery." They said they wanted to go to college and learn about computers, the future Verma was nudging them toward. They travel 18 miles daily just to attend grade school. It's a two-hour journey each way, by bicycle and bus.

I asked, "What do you hope to learn on computers?"

"We would like to get all the faraway information," said Bhawana. There was more talk in Hindi between Verma and the girls. I let it pass for a few minutes and finally asked, "What are you discussing now?"

"I am telling them that in America, everything is done by machines. The tea has been prepared by a machine. When they are extracting milk from the buffalo, it is done by machines."

Bhawana now asked me a question. "Are you going to feel happy after visiting this village?"

"Will I feel happy?"

I stuttered through something about the merits of living without machines, about the fact that I didn't know how to do anything because machines did everything for me, how we drank cow's milk, though I found India's buffalo milk delicious. Verma didn't even bother translating. He let me finish and turned away.

A few months earlier Verma had found Bhawana waiting outside an OMC plant in the rain. That is how they met. She stood on the brown road beneath a dark green bough, schoolbag in hand, surrounded by rain, waiting for someone to emerge from the plant who could explain this thing called the Internet. Verma showed her.

Soon Bhawana was using Verma's computer to look up market prices for sugarcane, mangoes, wheat, and rice so her father, a farmer, could use these as negotiating points. It was a tool for her father's agriculture business, as practical as a hoe. It fit into the farming lifestyle like the cell phones that enabled people to read weather forecasts, like the portable lanterns they carried into the fields at harvest time, when they worked all night pulling mangoes from big trees.

One of the Singhs brought out chai on a tray. It's something everyone does here. Chai, fried things, water with sugar mixed in —people always give you the best thing they have. And Verma took the moment to make sure I was noticing the world around me. "See that this is the village," he said. "See that their roofs are of grasses and all. See mud houses, a few brick. See that you can see the darkness. Only the moonlight is there."

BURKHARD BILGER

In Deep

FROM *The New Yorker*

ON HIS 13TH day underground, when he'd come to the edge of the known world and was preparing to pass beyond it, Marcin Gala placed a call to the surface. He'd traveled more than three miles through the earth by then, over stalagmites and boulder fields, cave-ins and vaulting galleries. He'd spidered down waterfalls, inched along crumbling ledges, and bellied through tunnels so tight that his back touched the roof with every breath. Now he stood at the shore of a small, dark pool under a dome of sulfurous flowstone. He felt the weight of the mountain above him—a mile of solid rock—and wondered if he'd ever find his way back again. It was his last chance to hear his wife's and daughter's voices before the cave swallowed him up.

"Base camp, base camp, base camp," he said. "This is Camp Four. Over." His voice traveled from the handset to a Teflon-coated wire that he had strung along the wall. It wound its way through sump and tunnel, up the stair-step passages of the Chevé system to a ragged cleft in a hillside 7,000 feet above sea level. There, in a cloud forest in the state of Oaxaca, Mexico, lay the staging area for an attempt to map the deepest cave in the world—a kind of Everest expedition turned upside down. Gala's voice fell soft and muffled in the mountain's belly, husky with fatigue. He asked his seven-year-old, Zuzia, how she liked the Pippi Longstocking book she'd been reading, and wondered what the weather was like on the surface. Then the voice of Bill Stone, the leader of the expedition, broke over the line. "We're counting on you guys," he said.

"This is a big day. Do your best, but don't do anything radical. Be brave, but not too brave."

Gala had been this deep in the cave once before, in 2009, but never beyond the pool. A baby-faced Pole of unremarkable physique—more plumber than mountaineer—he discovered caving as a young man in the Tatra Mountains, when they were one of the few places he could escape the strictures of communism. When he was 17, he and another caver became the first people to climb, from top to bottom, what was then the world's deepest cave, the Réseau Jean Bernard, in the French Alps. Now 38, he had explored caves throughout Europe and Ukraine, Hawaii, Central America, and New Guinea. In the off season, he was a technician on a Norwegian oil platform, dangling high above the North Sea to weld joints and replace rivets. He was not easily unnerved. Then again, Chevé was more than usually unnerving.

Caves are like living organisms, James Tabor wrote in *Blind Descent,* a book on Bill Stone's earlier expeditions. They have bloodstreams and respiratory systems, infections and infestations. They take in organic matter and digest it, flushing it slowly through their systems. Chevé feels more alive than most. Its tunnels lie along an uneasy fault line in the Sierra de Juárez mountains and seethe with more than seven feet of rain a year. On his first trip to Mexico, in 2001, Gala nearly died of histoplasmosis, a fungal infection acquired from the bat guano that lined the upper reaches of a nearby cave. The local villagers had learned to steer clear of such places. They told stories of a malignant spirit that wandered Chevé's tunnels, its feet pointing backward as it walked. When Western cavers first discovered the system, in 1986, they found some delicate white bones beneath a stone slab near the entrance: the remains of children probably sacrificed there hundreds of years ago by the Cuicatec people.

When the call to base camp was over, Gala hiked to the edge of the pool with his partner, the British cave diver Phil Short, and they put on their scuba rebreathers, masks, and fins. They'd spent the past two days on a platform suspended above another sump, rebuilding their gear. Many of the parts had been cracked or contaminated on the way down, so the two men took their time, cleaning each piece and cannibalizing components from an extra kit, knowing that they'd soon have no time to spare. The water here was between 50 and 60 degrees—cold enough to chill you within minutes—and Gala had no idea where the pool would lead. It might offer swift passage to the next shaft or lead into an endless, mud-dimmed labyrinth.

The rebreathers were good for four hours underwater, longer in a pinch. They removed carbon dioxide from a diver's breath by passing it through canisters of soda lime, then recirculating it back to the mouthpiece with a fresh puff of oxygen. Gala and Short were expert at managing dive time, but in the background another clock was always ticking. The team had arrived in February, three months before the rainy season. It was only mid-March now, but the weather wasn't always predictable. In 2009 a flash flood had trapped two of Gala's teammates in these tunnels for five days, unsure if the water would ever recede.

Gala had seen traces of its passage on the way down: old ropes shredded to fiber, phone lines stripped of insulation. When the heavy rain began to fall, it would flood this cave completely, trickling down from all over the mountain, gathering in ever-widening branches, dislodging boulders and carving new tunnels till it poured from the mountain into the Santo Domingo River. "You don't want to be there when that happens," Stone said. "There is no rescue, period." To climb straight back to the surface, without stopping to rig ropes and phone wire, would take them four days. It took three days to get back from the moon.

The truth is they had nowhere better to go. All the pleasant places had already been found. The sunlit glades and secluded coves, phosphorescent lagoons and susurrating groves had been mapped and surveyed, extolled in guidebooks, and posted with Latin names. To find something truly new on the planet, something no human had ever seen, you had to go deep underground or underwater. They were doing both.

Caving is both the oldest of pastimes and the most uncertain. It's a game played in the dark on an invisible field. Until climbing gear was developed, in the late 19th century, a steep shaft could end an expedition, as could a flooded tunnel—cavers call them terminal sumps. If an entrance wasn't too small or a tunnel too tight, the cave could be too deep to be searched by torch or candlelight. In the classic French caving books of the 1930s and '40s, *Ten Years Under the Earth,* by Norbert Casteret, and *Subterranean Climbers,* by Pierre Chevalier, the expeditions are framed as manly jaunts belowground—a bit of stiff exercise before the lapin chasseur back at the inn. The men wear oilskins and duck-cloth trousers, carry rucksacks and rope ladders, and light their way with a horse-carriage lantern. At one point in *Subterranean Climbers,* a sweet scent of Chartreuse fills the air and the party realizes, with dismay, that their digestif has come to grief against a fissure wall. Later a rock tumbles loose from a shaft and conks a caver named François on the head, causing some discomfort. The victim, Chevalier notes with regret, was "poorly protected by just an ordinary beret."

Chevalier and his team went on to map more than 10 miles of caverns in the Dent de Crolles, outside Grenoble. Along the way they set the world depth record—2,159 feet—and developed a number of caving tools still used today, including nylon ropes and mechanical ascenders. Casteret may have done even more to transform the sport. In the summer of 1922 he was hiking in the French Pyrenees when he noticed a small stream flowing from the base of a mountain. He shucked off his clothes and lit a candle, then wedged himself through the crack and waded in. The tunnel followed the stream for a couple of hundred feet, then dipped below the waterline. Rather than turn back, Casteret set his candle on a ledge, took a deep breath, and swam ahead, groping the wall till he felt the ceiling open up above him. He went on to explore many miles of tunnels inside the cave, culminating in a pair of large, airy galleries. The first was covered in spectacular limestone formations. The second was smaller and drier, with a dirt floor. When Casteret held his candle up to its walls, the flame flickered over engravings of mammoths, bison, hyenas, and other prehistoric beasts—the remains of a religious sanctuary some 20,000 years old.

Casteret and Chevalier helped turn caving into a heroic undertaking and the search for the world's deepest cave into an international competition—a precursor to the space race. "Praise Heaven, no one can give France lessons in this matter of epic achievement," Casteret wrote in his preface to Chevalier's book. "The race of explorers and adventure-seekers has not died out from our land." By combining lighter, stronger climbing gear with scuba tanks, cavers went deeper and deeper into the earth, more than tripling Chevalier's depth in the next 60 years. The record would bounce between France, Spain, and Austria (where one of

Gala's teams went below 5,300 feet at Lamprechtsofen in 1998) before settling in the Republic of Georgia in 2004.

A cave's depth is measured from the entrance down, no matter how high it is above sea level. When prospecting for deep systems, cavers start in mountains with thick layers of limestone deposited by ancient seas. Then they look for evidence of underground streams and for sinkholes—sometimes many miles square—where rain and runoff get funneled into the rock. As the water seeps in, carbon dioxide that it has picked up from the soil and the atmosphere dissolves the calcium carbonate in the stone, bubbling through it like water through a sponge. In Georgia's Krubera Cave, in the Western Caucasus, great chimneylike shafts plunge as much as 500 feet at a time, with crawl spaces and flooded tunnels between them. The current depth record was set there in 2012, when a Ukrainian caver named Gennadiy Samokhin descended more than 7,200 feet from the entrance—close to a mile and a half underground.

The Chevé system is even deeper. Drop some fluorescent dye into the stream at the entrance, as a teammate of Stone's did in 1990, and it will tumble into the Santo Domingo eight days later, 11 miles away and 8,500 feet below. No other cave in the world has such proven depth (though geologists suspect that some caves in China, New Guinea, and Turkey go even deeper). But that isn't enough to set a record: cave depths, unlike mountain heights, are inherently subjective. Everest was the world's tallest peak long before Edmund Hillary and Tenzing Norgay scaled it. But a cave is only officially a cave when people have passed through it. Until then it's just another hole in the ground.

Deep caves rarely call attention to themselves. Like speakeasies and opium dens, they tend to hide behind shabby entrances. A muddy rift will widen into a shaft, a crawl space into a vaulting nave. Krubera begins as a grave-size hole full of moss and crows' nests. When local cavers first explored it, in 1960, they got less than 300 feet down before the shaft leveled off into an impassable squeeze. It was more than 20 years before the passage was dug out, and another 17 before a side passage revealed the vast cave system beneath it. Yet the signs were there all along. The bigger the cave, the more air goes through it, and Krubera was like a wind tunnel in places. "If it blows, it goes," cavers say.

Chevé has what cavers call a Hollywood entrance: a gaping maw in the face of a cliff, like King Kong's lair on Skull Island. A long golden meadow leads up to it, bordered by rows of pines and a stream that murmurs in from the right. It feels ceremonial somehow, like the approach to an altar. As you walk beneath the overhang, the temperature drops, and a musty, fungal scent drifts up from the cave's throat, where the children's bones were found. The stream passes between piles of rubble and boulders, their shadows thrown into looming relief by your headlamp. Then the walls close in and the wind begins to rise. It's easy to see why the Cuicatec felt that some dark presence abided here—that something in this place needed to be appeased.

Like Krubera, Chevé starts with a precipitous drop: 3,000 feet in less than half a mile. But then it levels off to a more gradual slope: to go another vertical mile, you have to go 10 miles horizontally, at least half a mile of it underwater. Although the water eventually gathers into a single stream, the cave's upper reaches are full of oxbows and tributaries, meandering and intertwining through the rock, paralleling one another for a stretch, then veering apart or abruptly ending. It's tempting to imagine the system as a giant Habitrail, with cavers scurrying through it. But these tunnels weren't meant for inhabitants. They're geological formations, differentially eroded, their soft deposits ground down to serrated edges or carved into knobs and spikes that the body has to contort itself around. A long squirm down a tight shaft will lead to an even longer crawl, a slippery descent, and so on, in a natural obstacle course, relentless in its challenges. Near the main entrance there's a 30-foot section known as the Cat Walk, where a caver can hoist his pack and stroll forward without thinking. It's the only place like it in the system. "Every other piece of this cave might kill you," Gala told me.

Bill Stone has led seven expeditions to Chevé in the past 10 years, all but one of them with Gala. In 2003 his team dove through a sump that had thwarted cavers for more than a decade, then climbed down to nearly 5,000 feet, making Chevé the deepest cave in the Western Hemisphere. But there was no clear way forward: the main passage ended in a wall of boulders. The only option was to try to bypass the blockage by entering the system farther downslope. The following spring Stone sent teams of Polish, Spanish, Australian, and American cavers bushwhacking across the cloud forest in search of new entrances. They found more than 100, including a spectacular cliff-face opening called Atanasio. The most promising, though, was a more modest but gusty opening labeled J2 (the *J* was for *jaskinia,* Polish for "cave"). It was wide open at the top but pinched tight as soon as you went down. The Australians called it Barbie.

The J2 system runs roughly parallel to the main Chevé passage and about 1,000 feet above it. The water's exact course through the mountain is hard to predict, but cave surveys and Stone's 3-D models suggest that the two systems eventually merge. If Gala and Short could get past the sump beyond Camp Four, their route should join up with Chevé, drop another 2,500 feet, and barrel down to the Santo Domingo. "Imagine a storm-tunnel system in a city," Stone told me. "All these feeders connect to a trunk and then go out to an estuary. We're in the back door trying to get into that primary conduit." This is it, he said. This is the big one. "If everything goes well, we'll be as far as anyone has ever been inside the earth."

Deep caving demands what Stone calls siege logistics. It's not so much a matter of conquering a cave as outlasting it. Just to set up base camp in Mexico, his team had to move six truckloads of material more than 1,200 miles and up a mountain. Then the real work began. Exploring Chevé is like drilling a very deep hole. It can't be done in one pass. You have to go down a certain distance, return to the surface, then drill down a little farther, over and over, until you can go no deeper. While one group is recovering on the surface, the other is shuttling provisions farther into the cave. Stone's team had to establish four camps underground, each about a day's hike apart. Latrines had to be dug, ropes rigged, supplies consumed, and refuse carried back to the surface. Divers like Gala and Short were just advance scouts for the mud-spattered army behind them, lugging 30-pound rubber duffel bags through the cave—Sherpas of a sort, though they'd never set foot on a mountaintop. Stone called them mules.

Two months earlier, in Texas, I'd watched the final preparations for the trip. Stone's headquarters are about 15 minutes southeast of Austin, on 30 acres of drought-stricken scrub. There is a corrugated building out front that's home to Stone Aerospace, a robotics firm he started in 1998, and a two-story log house in back, where he lives with his wife, Vickie, a fellow caver. (They met at a party where Stone overheard her talking about tactical rigging.) The trucks were scheduled to leave in two days, and every corner of the house had been requisitioned for supplies. One room was piled with cook pots, cable ladders, nylon line, and long underwear. Another had dry suits, diving masks, rebreathers, and oxygen bottles. In the basement, eight long picnic tables were stacked with more than 1,000 pounds of provisions. Shrink-wrapped flats of peanuts, cashews, and energy bars sat next to rows of four-liter bottles filled with staples and dry mixes: quinoa, oatmeal, whey protein, mangoes, powdered potatoes, and broccoli-cheese soup. Stone had tamped in some of the ingredients using an ax handle.

"In the past I'd lose twenty-five pounds on one of these trips," Stone told me. "We can burn as many calories as a Tour de France rider every day underground." Ascending Chevé, he once said, was like climbing Yosemite's El Capitan at night through a freezing waterfall. To fine-tune the team's diet, he'd modeled it on Lance Armstrong's program, aiming for a ratio of 17 percent protein, 16 percent fat, and 67 percent carbohydrates. In Mexico the supplies would be replenished with local beans, vegetables, and dried machaca beef. "What you aren't going to find is candy," Stone said. "Stuff like Snickers—that's bullshit." When I looked closer, though, I found a bottle of miniature chocolates that Vickie had hidden among the supplies.

Cavers, even more than climbers, have to travel light and tight. Bulky packs are a torture to get through narrow fissures, and every ounce is extracted tenfold in sweat. Over the years caving gear has undergone a brutal Darwinian selection, lopping off redundant parts and vestigial limbs. Toothbrushes have lost their handles, forks a tine or two, packs their adjustable straps. Underwear is worn for weeks on end, the bacteria kept back by antibiotic silver and copper threads. Simple items are often best: Nalgene bottles, waterproof and unbreakable, have replaced all manner of fancier containers; cavers even stuff their sleeping bags into them. Yet the biggest weight savings have come from more sophisticated gear. Stone has a PhD in structural engineering from the University of Texas and spent 24 years at the National Institute of Standards and Technology, in Gaithersburg, Maryland. His company has worked on numerous robotics projects for NASA, including autonomous submarines destined for Europa, Jupiter's sixth moon.

The rebreathers for the Chevé trip were of his own design. Their carbon-fiber tanks weighed a fourth of what conventional tanks weigh and lasted more than four times longer underwater; their software could precisely regulate the mix and flow of gases.

Stone's newest obsession was a set of methanol fuel cells from a company called SFC Energy. Headlamps, phones, scuba computers, and hammer drills (used to drive rope anchors into the rock) all use lithium batteries that have to be recharged. On this trip the cavers would also be carrying GoPro video cameras for a documentary that would be shown on the Discovery Channel. In the past Stone had tried installing a paddle wheel underground to generate electricity from the stream flow, with fairly feeble results. But a single bottle of methanol and four fuel cells, each about the size of a large toaster, could power the whole expedition. The question was whether they'd survive. High-tech gear tends to be fragile and finicky. While I was in Texas, one of the rebreathers kept shutting down for no apparent reason (it was later found to have a faulty fail-safe program), and this was the sixth generation of that design. The fuel cells weren't nearly as robust. Stone would keep them in shockproof, watertight cases, but he doubted that would suffice. "We're going to take them down there and turn them into broken pieces of plastic," he said.

Stone knew what it meant to be a battered piece of hardware: he'd turned 60 that December and had spent more than a year of his life underground. His gangly frame—six feet four, with a wingspan nearly as wide—was kept knotty by free weights, and he could still outclimb and outcarry most 25-year-olds. But he was getting old for an extreme sport like this, and he knew it. He had the whiskered, weather-beaten look of an old lobsterman. "I think it's a little surprising to him how hard the caving is on his body these days," one of the team members told me. "I won't say that he's feeling his age, but he's realizing that he isn't at the pointy end of the stick anymore."

As a leader Stone models himself on the great expeditionary Brits of the past century. He has an engineer's methodical mind and an explorer's heroic self-image. He's pragmatic about details and romantic about goals. His teammates often compare him to Ernest Shackleton, another explorer who felt most alive in the world's most unpleasant places. But Shackleton, despite shipwreck and starvation, never lost a man under his direct command. ("I thought you'd rather have a live donkey than a dead lion," he told his wife after failing to reach the South Pole.) Cave diving is less forgiving. Stone has lost four teammates on his expeditions, including Henry Kendall, the Nobel Prize–winning physicist. Kendall failed to turn on the oxygen in his rebreather while cave diving in Florida. Others have succumbed to narcosis or hypoxia, fallen from cliffs or had grand mal seizures, lost their way or lost track of time. They've buried themselves so deep that they couldn't come back up.

Stone's single-minded, almost mechanistic style can sometimes raise hackles. He can be inspiring one moment and dismissive the next. "Bill has problems identifying people's emotions," Gala told me. "So he doesn't always react to them well." Then again, it's hard to avoid tension in a sport that takes such a mortal toll. Stone's mentor, the legendary cave diver Sheck Exley, retrieved 40 corpses from diving sites in Florida alone, then drowned in a Mexican cenote in 1994. "When cavers become cave divers, they usually die because of it," Stone's friend James Brown told me. In 1988 Brown and Stone were called in to help remove the body of a female diver from a cave near Altoona, Pennsylvania. When they found her, she was tangled in rope at the bottom of a sump, arms so stiff that, Brown recalled, Stone suggested they cut them off for easier transport. "Nobody liked that idea much," Brown said. "But after a while her arms softened up, and we were able to fold them down."

It took them two days to get her out, with Stone pushing from behind. "He kept saying, 'Don't leave me back here if she gets stuck!'" Brown said. If there's one rule of caving, Stone told me, it's that you never leave a person behind. Especially if they're alive, he added. "If they're dead, it's another matter."

By the time I arrived at base camp, in mid-March, the team had settled into a soggy routine. A week underground followed by 10 days on the surface. Five days of drizzle followed by one day of sun. They'd spent most of the first month hauling gear up the mountain—a muddy three-hour hike from a farmhouse in the valley—loading the heaviest items on burros and the rest on their backs. They'd set up tents and dug latrines, strung lights and cut trails to the cave. The camp was spread out beneath pines and low-hanging clouds, on a rare stretch of relatively flat ground. To one side the Discovery crew had erected a geodesic dome with two full editing stations inside. To the other the cavers had hung a giant blue tarp, sheltering a long plywood table, stacks of provisions, and a pair of two-burner camp stoves. On most expeditions base camp is a place to dry out and recover from infections acquired underground—cracked skin and inflamed cuts and staph bacteria that burrow under your fingernails till they ooze pus. But this forest was nearly as wet as the cave.

"Welcome to hell," one of the cavers told me when I joined him by the campfire that first night. "Where happiness goes to die," another added. There was a pause, then someone launched into the colonel's monologue from *Avatar:* "Out there, beyond that fence, every living thing that crawls, flies, or squats in the mud wants to kill you and eat your eyes for jujubes . . . If you wish to survive, you need to cultivate a strong mental attitude." It was a favorite conceit around camp: the cloud forest as hostile planet. But looking at all the gleaming eyes around the fire, I was mostly reminded of the Island of Lost Boys. Beneath all the mud and gloom and dire admonitions, there burned an ember of self-satisfaction—of pride in their wretched circumstance and willingness to endure it. As Gala put it, "It's just one continuous miserable."

Fifty-four cavers from 13 countries, 43 of them men and 11 women, would pass through the camp that spring. The team had a core of 20 or so veteran members, reinforced by recruits from caving groups worldwide. On any given day the cave might be home to a particle physicist from Berkeley, a molecular biologist from Russia, a spacecraft engineer from Washington, D.C., a rancher from Mexico, a geologist from Sweden, a tree surgeon from Colorado, a mathematician from Slovenia, a theater director from Poland, and a cave guide from Canada who lived in a Jeep and spent 200 days a year underground. They were a paradoxical breed: restlessly active yet fond of tight places, highly analytical yet indifferent to risk. They seemed built for solitude—pale, phlegmatic creatures drawn to deep holes and dark passages—yet they worked together as a selfless unit: the naked mole rats of extreme sport. As far as I could tell, only two things truly connected them: a love of the unknown and a tolerance for pain.

Matt Covington, a 33-year-old caver from Fayetteville, was a typical specimen. A professor of geology at the University of Arkansas, he had earned his PhD in astrophysics but switched fields so that he could spend more time underground. He had a build best described as Flat Stanley. Six feet four but only 150 pounds, he could squeeze through a crevice six and a half inches wide. "My head isn't the limiting factor," he told me. "It's my hips." Covington was a veteran of seven Stone expeditions as well as caving trips to Sumatra, Peru, and other remote formations. Five years earlier he was climbing up a cliff face in Lechuguilla Cave, near Carlsbad Caverns, when an anchor came loose from the rock. Covington's feet caught on the cliff as he fell, tumbling him onto his left arm, causing compound fractures. Rather than wait for rescue, he spent the next 13 hours dragging himself to the surface. "The crawling was fairly uncomfortable," he allowed. "There was a lot of rope to climb."

When I first met Covington, late one night, he'd just slouched back into camp after five days underground. His eyes were bloodshot, his blond hair clumped and matted, his skin as blanched and fuzzy as moldy yogurt. He was so tired that he could barely stand, and his clothes reeked of cave funk. Yet he seemed fairly content. "A good caver is one who forgets how bad it really is," he said. There was more to it than that, though. Covington didn't feel claustrophobic underground; he felt at home. The rock walls, to him, offered a kind of embrace. As a boy, he told me, he used to flop around so much in his sleep that he often fell on the floor. Rather than climb back up, he'd crawl under the bed and stay till morning. He felt better there, beneath the springs, than he did looking up at the ceiling in his big empty room.

It was an instinct almost everyone here seemed to share. One of the cavers remembered staring at a slice of rye bread as a child, fascinated by all the air bubbles beneath the crust. He wanted to go down there. Gala was so comfortable in caves that he sometimes felt as if they were made for humans. "The passages are exactly the right size for my body to fit in," he told me. And his wife, Kasia, who worked as a photo editor in Warsaw, was nearly as happy underground as he. They took turns exploring the cave and taking care of their daughter, Zuzia, up on the surface. Zuzia had spent much of her life watching people disappear into holes and reemerge weeks later. She traversed her first cliff face at the age of four, in Spain's Picos de Europa mountains, and kept a map above her bed with pirate flags pinned on all the countries she'd visited. When she first came to Mexico, in 2009, she would sometimes cry out in frustration, "It's so uncomfortable here!" Now she flitted between tents like a forest sprite, half naked in the cold, fencing with corncobs and setting traps for mice. Life at camp had built up her immune system, Gala assured me, and had taught her the "skills of dynamic risk assessment."

I wished that I could see Chevé through her eyes. Before her father went underground with Phil Short, for their long hike beyond Camp Four, he'd read to Zuzia from *The Hobbit.* Chevé was no Lonely Mountain. Yet it had glistening caverns and plummeting boreholes, stalagmites tall as organ pipes and great galleries draped in flowstone, deeper than any goblin lair. And they were right beneath her feet. "When you squeeze through these small holes into these big halls, you feel like you're the only person on the earth," Gala said. "It's like the kingdom of the dwarves."

Gala had been exploring Chevé with Stone so long that he could nearly navigate it blindfolded. After a while, he said, you start to create a map of the system in your mind, to memorize each contortion and foothold needed to climb through a passage. On the steepest pitches, certain rocks almost seemed to smile and wave at him and to reach for his hand. He would grab them, thinking, Old friend! And yet the deeper he went, the more unfamiliar the territory became. By the 13th day, the escalating uncertainty—the risk of a careless stumble or a snapped limb so far from the surface—was starting to weigh on him. "The further in you go, the more you begin to doubt and question yourself," he told me. "What the fuck am I doing here?"

The sump beyond Camp Four was like nothing Short and Gala had seen before. The three sumps higher up in this system were relatively shallow and less than 500 feet long. This sump was more than 30 feet deep, and it seemed to go on and on. And something more rare: it was beautiful. The water was a luminous turquoise, flowing over pure white sand; the limestone was streaked with ocher and rust. Most sumps are cloudy, tubelike passages carved by underground streams, but this one had been a dry cave not long ago. The stalactites on its ceiling could only have been formed by slow drip. With its lofty chambers and limpid water, it reminded Gala of the blue holes of Florida and the caves of the Yucatán. Finning through it felt like flying.

The hazards of cave diving are inseparable from its seductions.

Wide-open tunnels can fork into a maze; white sands swirl up to obscure your view. You think that you know the way back, only to reach a dead end, with no place to come up for air. "People think that cave diving is an adrenaline sport, but really it's the opposite," Short told me. "Whenever you feel your adrenaline racing, you have to slow down. Stop, breathe, think, act, and, in general, abort. That's the rule in cave diving."

Short is one of the sport's premier practitioners, with experience as far afield as the Sahara and shipwrecks off Guam. His body is a testament to its rigors: long and arachnid, skin taut over bone, head shaved to shed its last encumbrance. With his rapid-fire talk and glasses that seem to magnify his eyes, he could pass for a street preacher or a pamphleteer. But his absurdist wit was a great gift around a campfire, and his diplomacy often took the edge off Stone's blunt directives. Gala and Short were a good match: one quiet, the other loquacious; one expert at climbing, the other at diving. Just as Gala could pick his way through Chevé by memory and internal gyroscope, Short could divine a sump's path from half-conscious clues: the flow of current and its fluctuating temperature, the shape of the walls and ripples in the sand. Still, he took no chances. As they swam from chamber to chamber, the beams of their headlamps needling the dark, he unspooled a three-millimeter line behind him, like Theseus in the Labyrinth.

An hour later he signaled for Gala to stop. Below them in the sand was the line they'd laid down 15 minutes earlier. The tunnel had led them on a loop. They'd expected the sump to be about 1,000 feet long, but they'd already gone twice that distance, and time was running out. Cave divers like to ration their air supply by a rule of thirds: one part for the way in, another for the way out, and a third in reserve. On a four-hour rebreather, that left them less than half an hour for exploring. The cave was a honeycomb, they realized, with tunnels angling off in every direction and hardly any current to guide them. "There were passages everywhere, everywhere," Gala recalled. "It was so complex we could spend a year looking."

In the end they just picked a tunnel and hoped for the best. When they'd backtracked around the loop, reeling in their line, they came to a kind of four-way intersection. One passage led back to the beginning of the sump, another to the loop behind them, a third to a dead end they'd explored earlier. That left one unexplored passage. It took them up a short corridor, along a rising slope of terraced mustard-colored flowstone, and into a small domed chamber. There was an air bell at the top about the size of a car trunk, so they swam up and took off their helmets and neoprene hoods to talk. They seemed to be at a dead end. They were cold, tired, and disoriented, and their air ration had nearly run out. There was no choice but to head back. "We were just a little overwhelmed by this dive," Gala told me. Then they heard the waterfall.

A mile above them, at base camp, Stone was waiting impatiently for their call. This was the pivotal moment in the expedition—the day for which he'd spent four years perfecting gear, recruiting cavers, and raising money. (The total budget for the trip was roughly $350,000, most of it paid for by equipment sponsors and the Discovery Channel.) He had expected Gala and Short's reconnaissance trip to take less than six hours—two hours to dive the sump, two hours to look around and find a campsite, and another two to swim back and call in—yet nine hours had passed. "There are a bunch of scenarios that could be going on right now," he told the Discovery cinematographer, Zachary Fink. "Even a one-kilometer swim with fins would take only about an hour. And that was way beyond our limit."

Stone looked haggard and thin, his mustache drooping over sallow skin. Weeks of shuttling supplies into the cave had taken a toll on him. He was a strong climber and diver, but he wasn't a "squeeze freak" like some of the others. His broad, bony shoulders weren't built for these tunnels. In the tightest fissures he had to take off his helmet just to turn his head, or strip down to his dry suit and wriggle between walls for hundreds of feet. (They called one passage the Contusion Tubes.) "It's hypothermic as hell down there," he told me. "The wind is whipping through, the water's in contact with the rock, and you can just feel the calories being sucked out. It can be more dangerous than a high-altitude peak at twenty-five below." By the time he'd resurfaced a few days earlier, he was coming down with a flu. Then it rained for three and a half days.

It was late evening when the call finally came: "Base camp, base camp, base camp!" Stone rushed over to the phone and hit the talk button. "Tell us what happened," he said. There was a blast of static, then Short's clipped British accent came crackling over the line. "We have good news and we have complicated news," he said. "From a point of view of future exploration, complicated is today's understatement."

The waterfall could mean only one thing, Short and Gala knew. They'd reached the end of the sump and the river was flowing nearby. How to get there? When Gala ducked his head underwater and looked around, the chamber looked sealed off. But when he looked again his headlamp picked up an odd texture in the wall to his right. There was a gap in it just below the waterline—wide enough for a person to squeeze through. Gala could tell that his rebreather wouldn't fit, so he handed it to Short, along with his mask, helmet, and side tanks. "I left him holding all these things with his teeth and both his hands," he recalled later. Then he held his breath and dove through.

When he resurfaced on the other side, he was in a fast-flowing canal of clear water. The walls were formed by ancient breakdown piles, their boulders napped in calcite; the low ceiling was hung with stalactites. As he swam, a wide, airy passage opened up ahead, with a large pool in the distance. It glimmered in his headlight. He hiked over to it and swam across, feeling light and buoyant without his rebreather. He could hear the roar of the waterfall growing louder as he went, but an enormous stalagmite blocked the way, with only a thin gap to one side. He stretched an arm and a leg through the opening and shimmied around, thankful again to be rid of his gear. When he was through, he found himself in a great chamber filled with mist and spray, its floor split by a yawning chasm. The river ran into it from the right and fell farther than his light could follow. Across the chamber, 30 or 40 feet away, a huge borehole stretched into the darkness. This is it, Gala thought, the breakthrough they'd imagined. With any luck, it would take them straight to Chevé's main passage.

Stone wasn't so sure. "Is there any place at all over there that you saw that would be suitable for a camp?" he asked Short over the phone when the story was done. "Negative. There is not a single flat surface other than the surface of the river." Stone clutched his head and frowned. The sump was too long. Two thousand feet! They didn't have enough line down there to rig that distance.

Without rigging, most of the team couldn't dive the sump safely, and without their help Gala and Short couldn't resupply for the next push. "The whole game had changed," Stone told me later. "Just diving through wasn't the game. The game was to get all the support material to the other side. It was like running a war: if you don't get the food and fuel and ammo to the front line, you're going to stall out."

Only a few dozen people in the world had both the caving expertise and the scuba skills to go this deep in the cave. Of those, 12 had originally agreed to join the expedition. Then the number began to drop. Three died before the expedition began: one on a deep dive in Ireland, another in an underwater crevice in Australia, the third from carbon monoxide poisoning in Cozumel. Three had left early or had not yet arrived. And three had physical limitations: James Brown had gimpy knees, a Mexican diver named Nico Escamilla had a pulled groin muscle, and a veteran diver named Tom Morris had torn a rotator cuff. "It was like getting hit in the head with a two-by-four," Stone told me. "Oh, crap! We've lost most of our divers! The three that are qualified to dive the sump are the two that are down there and me—and, God bless them, Phil and Marcin want to see daylight."

It was too late to recruit new divers to the team. The best candidate, a veteran British caver named Jason Mallinson, had joined another expedition, across the river at a cave system called Huautla. "He's one of the best divers in the world," Stone told me. "But he has a certain personality—it's abrasive, and what I really wanted this year was harmony, and I got it." Stone had planned to join Gala and Short for the last leg of the expedition, to see the very deepest regions of the cave. But without more divers to support them he wasn't sure it was safe to continue. "They did a fantastic thing there, but it may also be the end of that route," he said after Short got off the line. "There is no glory in rushing into something like that and losing a friend. It just is not worth it."

Word of Stone's misgivings filtered down to Gala and Short as they worked their way back up the cave, camping with the support crews. It seemed a kind of betrayal. The yo-yo logistics of deep caving required that they return to the surface to rest and reprovision, but they had every intention of going back down. Yes, the sump was longer than expected, the conditions more challenging.

But they'd found exactly what they wanted on the other side. How could they stop now?

"My thinking was that Bill is just tired with this cave—that this is just an excuse not to come back," Gala told me. "I think that he spent too much time preparing this expedition, making all these tools, all these deals." But Stone insists that his reluctance was just a matter of safety and logistics—an equation like any other, balancing risk and reward. On Gala and Short's first evening back at base camp, the scene around the campfire got so tense that Stone shouted at Zachary Fink to turn off his camera. "It's always like that at some point in an expedition," Gala told me. "There's always a shouting match between Bill and me, with someone almost crying." But over bourbon that night and coffee the next morning, they slowly hashed out a plan. They would have to work fast, resupplying the camps themselves and exploring the new tunnels without backup divers. If they hung hammocks from the wall beyond Sump Four, they could bivouac there and explore the cave for another three weeks before they ran out of rope. With any luck, they'd reach the Chevé juncture before they were done.

Stone went underground the next day. Short took five days to rest and heal—half the usual recovery time, after three times the usual stay underground—and by the morning of March 21st he was leading a ragtag team down the mountain. This was just a five-day trip to help prepare the cave for the final push. But with the expedition so undermanned, Short had no choice but to lead the team and to bring two novices along: Patrick van den Berg, a hulking information-security specialist from Holland, and David Rickel, an emergency medical technician from Texas. Van den Berg was a weekend caver in relatively poor shape ("I get most of my exercise moving a mouse around," he told me). Rickel was the team medic. He had a rock climber's ropy build, but the closest he'd come to deep caving was working in an iron-ore mine in Australia.

Short was of two minds about taking them. He knew that one injury could derail the whole expedition and that the cave ahead would test even the fittest athlete. "You can lift weights and go wall climbing and run a few miles every day, but it's not the same," he told me as we wound our way down the slope. "When you're nineteen days underground, in the cold and wet without a bed, with a forty-pound pack on your back, crawling on your hands and knees or climbing up and down cliffs or diving through sumps, and then you come back and resurface, and four days into your ten-day break some sadist wants to send you back down under, and you end up volunteering to go—most people hear that and they think you're stark raving mad." Yet Short was an optimist at heart and an experienced teacher—he gave scuba and cave-diving lessons in England—and he'd seen even novices accomplish unimaginable things. "It's not the body that breaks, it's the mind," he said. "If you compare this to what the British infantry were lugging in the Ardennes in World War I, or what Shackleton's team did in the Antarctic, this shit is easy. They were trudging up those slopes with old-fashioned ropes and no oxygen, and I'm sitting here complaining about the hole in my antibacterial underwear."

Whether van den Berg took any comfort in this wasn't clear. Less than an hour from camp, he was already red-faced and wheezing, sweat streaming down his chest. The altitude was getting to him, he told me. Hiking at 7,000 feet made him feel like he was breathing through a straw. When the team set down its packs for a brief rest, Short came over and crouched beside him. "I'm a little concerned that you're as tired as you are after just walking down a hill," he told him. "Your pulse was up to one-sixty, which David tells me is pretty high." Van den Berg shook his head and insisted that he was fine. He wouldn't have a problem going down the cave. "But you have to come back out, too," Short told him.

We were headed toward a cave entrance known as the Last Bash, about a mile from base camp. Discovered in 2005, it was a side entrance to the J2 passage, an hour or so down the slope from the original entrance. It would allow the team to bypass a sump and cut 12 hours out of the trip down, but it was tighter and more punishing than the other entrance—just a crack in the rock 10 feet above the trail, flanked by boulders and elephant-ear vines. If Short hadn't pointed it out, I would have passed right by it.

Short's team peered up at the opening for a moment, then slowly put on their gear. They stepped into their waterproof caving suits and climbing harnesses, attached special ratchets for rappelling down cliffs, and strapped on their helmets and headlamps. "This is not going to be some macho-driven bullshit," Short assured them. "It's going to be a slow bumble down the cave, with double dinners when we get to camp." They made a quick snack of crackers and energy bars, while Rickel checked van den Berg's heart rate again. It had dropped to 120. "How did you end up here?" van den Berg asked him when he'd finished. Rickel laughed. "A long sequence of poor life choices," he said. Then they crawled into the crack one by one and disappeared.

The rains were getting to be a serious concern now. The tunnels below the Last Bash weren't known to flood, but neither were the tunnels above it before 2009. Then some gravel got clogged in a fissure at the bottom of a pool, flooding the chamber behind it and trapping seven people in the cave below. Five were able to dive out, but the other two, Nikki Green and David Ochel, had to sit and wait, not knowing if the tunnel would clear. "We had no food for five days, just watching the water," Green told me. In the end the rain abated long enough for them to climb out, and then the cave flooded for the rest of the season. "We stayed too long," Green said.

A neighboring cave, known as Charco, had an even more unpleasant history. In 2001 a team of six cavers was heading back to the surface there after a week of surveying when they noticed the underground stream starting to rise. It had been raining for two days by then, and the tunnel was so tight that it began to flood. Charco is a place to make even cavers claustrophobic: the first camp is a 12-hour crawl from the surface, mostly on your belly. By the time the last team member neared the entrance, the water in the tunnel was inches from the ceiling. As he treaded water, lifting his face up to breathe, bits of soft white debris drifted toward his mouth and got caught in his hair. But it wasn't debris, as it turned out. A cow had died in the entrance that spring. Its belly was infested with maggots and the rains had washed them into the cave.

If there was an advantage to going deep, it was that the cave was fairly sterile. In the lower reaches of J2, the only signs of life were a few translucent crustaceans and bits of refuse that washed down from above. (In the Huautla system, teams sometimes found Popsicle sticks a mile belowground.) By early April the camps were reprovisioned, Rickel and van den Berg were safely back on the surface with Stone, and Gala and Short were alone once again at Camp Four. The sump beyond it, once the dark side of the moon, now seemed comfortingly familiar. Short had discovered a larger opening in the chamber at the end, which allowed them to dive out with their rebreathers and equipment. When they had swum down the canal on the other side and followed the tunnel to the misty chamber with the waterfall, it was as if they'd arrived at another beginning. "Now we were in a truly dry, unexplored cave," Short told me. "Our lights were the first light that had fallen on this place since it had been created."

Two promising passages lay ahead: the fossil gallery where the river had once flowed, and the canyonlike fissure where it now fell. They took a moment to gather themselves at the top of the falls and to make a pot of hot chocolate. But Gala couldn't bear to wait. While Short tended the stove, he free-climbed the 40 feet to the other side of the canyon—ropes would come later. Then he shouted at Short to join him.

It was just as they'd hoped: a cavernous passage, perhaps 15 feet high by 30 feet wide, with a packed mud floor. There was even a flat spot ahead where they could set up a camp. The gallery followed the path of the tunnel behind them at first, then meandered left and right, up and down. Gala and Short took surveyor's notes as they went, one man walking ahead and holding up a saucepan lid while the other shot a laser at it to get the distance. They used a compass and a clinometer to measure the tunnel's direction and slope, marked the numbers with a Sharpie onto a waterproof sheet, then copied them onto a piece of colored tape and tied the tape to the reference point. (Back at base camp, Stone would enter the data on his laptop to create 3-D maps of the cave.) This was standard practice in new tunnels and could add hours to a trip. But not here: after 300 or 400 feet, the passage abruptly ended. Rather than drop down to rejoin the stream, it had circled back on itself like the oxbow in the sump, ending in a large chamber walled with flowstone. It would take them no farther.

Gala and Short trudged back the way they'd come, their spirits deflated. A dry fossil gallery is the caver's version of a superhighway: the fastest, safest way underground. But at least they had another option. "There was still the waterfall," Gala told me, "and it *had* to go further down." He and Short strapped on their climbing harnesses and unpacked their rigging. The hammer drill had gone dead after the battery got wet—the fuel cells had all met the same fate—so Gala had to knot the rope around a rock to anchor it. But it held firm as they rappelled down the chasm. Forty feet below, the water thundered into a shallow pool, then slipped down a stair-step streambed to another, much larger pool below. They'd left their dry suits at the top of the falls to air out, so they had no choice but to swim across in their thermal underwear. The water here was a few degrees warmer than higher up in the cave but still close to 40 degrees below body temperature, and the sopping cloth kept it close to their skin. Yet they kept moving forward. "Expedition fever had bitten us," Short says.

When they reached the far shore, the water cascaded down to yet another pool, 20 feet below. They rigged ropes for the descent, scrabbled down, and swam across, their limbs trembling as the cold sank into them. In the distance the dusty beam of Gala's headlamp picked out a pile of boulders in their path, but this only quickened his pulse. It reminded him so clearly of a passage higher up, where a series of pools led to a breakdown pile along a fault line and then a wide-open tunnel beyond it. "I had this feeling that we were almost done," he told me. "We will climb these boulders. We will find a huge borehole, and that will open the way to Chevé."

It was not to be. When Gala and Short arrived at the breakdown pile, it was just the back end of a small sealed chamber—another cul-de-sac. Its boulders were bound together with flowstone, the holes between them no larger than your hand. "There was no air, no anything," Gala recalls. As for the river, it had found a long crack in the floor less than an inch wide, and spooled through it like an endless bolt of turquoise cloth.

They stood there for a moment in shock, not quite believing that they'd reached the end. They knew that the cave kept on going below, gathering the waters of Chevé beneath them. Yet there was no way forward. Like the cavers in Krubera before the side tunnel was discovered, they had yet to unlock the system's secret door. Gala looked over at Short—he was shaking uncontrollably now, his wiry limbs lacking all insulation—and was grateful once again to have him at his side. "It's like a friendship during war," he told me. "So strong an experience, it ties souls together." He clasped Short's shoulder and told him to go make some hot drinks while he finished surveying. Then they packed up their gear and began the long climb back to the surface.

Deep caving has no end. Every depth record is provisional, every barrier a false conclusion. Every cave system is a jigsaw puzzle, groped at blindly in the dark. A mountain climber can at least pretend to some mastery over the planet. But cavers know better. When they're done, no windy overlook awaits them, no sea of salmon-tinted clouds. Just a blank wall or an impassable sump and the knowledge that there are tunnels upon tunnels beyond it. The earth goes on without them. "People often misunderstand," Short told me. "All you find is cave. There is nothing else down there."

When I spoke to Stone recently, he was already planning his next trip to Chevé. His team had brought back some intriguing data, he said. Gala's survey showed that the end of J2 lies directly below a cave entrance discovered in the early '90s. The tunnel beyond it is fairly cramped, but there's enough air blowing through to suggest that it leads to a larger passage—one that could bypass the blockage in J2. If Stone's team can connect the two tunnels, then drop down into the main Chevé passage, they might still stitch the whole system together. "Where did the water go a million years ago? That's what you have to ask yourself," Stone said. "As a cave diver, you have to think four-dimensionally." In the meantime, this spring he was joining an expedition across the river to Huautla, where Jason Mallinson had managed to reach a depth of more than 5,000 feet—a new record for the Western Hemisphere. Huautla can never go as deep as Krubera, Stone said, much less the full Chevé system. But it could well be the *longest* deep cave in the world. Why not see how far it goes?

That was as good a reason as any. For most of the team, though, it wasn't the chance at a record that would bring them back, or even the lure of virgin cave. It was the camaraderie underground—the deep fellowship of shared misery. The camps down there were just a few damp tents on rubble, clustered around a propane flame. The food was the same dehydrated stuff they ate up top. A trip to the latrine could be a life-threatening experience—a squat on slippery rocks above a thundering chasm. But after weeks underground, even that smell could lift your spirits. It held the promise of dry clothes and hot coffee, black humor and noisy sex, drowned out by sing-alongs. Gala and Short spent one very good night hollering "C Is for Cookie" until they were hoarse.

On their 21st day underground, when they finally emerged from the cave's rocky clutch, they blinked up at the sun like newborns. Their skin was ashen, their eyes owl-wide and dilated. "I had these mixed emotions," Gala told me. "I understood that this is the end of J2—nine years of my life, of the most beautiful exploration of my life. It was a sad story." Yet it had also been the longest and hardest trip he'd ever taken, and it made the return to the surface all the sweeter. The green of the forest, so luminous and deep, seemed nearly psychedelic after weeks of dun-colored earth and the pale wash of his headlamp. The smell of leaves and rain and the workings of sunlight were almost overwhelming.

"It is beautiful here, isn't it?" Gala had told me when we first met, on a gray, drizzly morning at base camp. "Listen to these strange birds! When I'm back on the surface, just by contrast, I enjoy every piece of my life. Everything is fantastic." He laughed. "Some people say that all this caving is just for a better taste of tea."

SHEILA WEBSTER BONEHAM

A Question of Corvids

FROM *Prime Number Magazine*

1. *Corvidae* Corvus brachyrhynchos

> If men had wings and bore black feathers,
> few of them would be clever enough to be crows.
> —Henry Ward Beecher

Birds are everywhere here on the Carolina coast. Pelicans skim the blue-bellied rollers, bank for advantage, plummet and rise. Sanderlings and stilts drill for morsels in the sand while egrets stalk the marshes. Birds are everywhere. They are hungry, and they come to dine on the veranda of this inn on the beach. Flocks of gulls hang heavy-bodied over the tables long enough to check for an unguarded bit of fish or bread or meat, and the bold ones find the thing they all want. I watched a herring gull last October pluck a fillet from between two halves of a tourist's bun and rise on the same wingbeat. Laughing gulls, with their black bonnets and chuckling calls, are less common, but they do come, mostly in spring. Other birds, too. Pigeons, of course. Grackles and cowbirds. Dozens of the hard-to-name wee guys that birders call LBJs —"little brown jobs"—flit here and scurry there for crumbs and handouts.

Crows. If gulls are the berserkers of birdkind, swooping and screaming and plundering, then corvids, including crows, are the strategists. They watch. Face a crow at close quarters and you see that *you* are the one under study. With an eye sharper than his pointed bill, the crow pins down your moves and knows you better than you know him. Scientists have documented what farmers have said through the ages: crows can count. They communicate. Consummate mimics, they even copy human speech.

Picture this: you are sitting on the hotel veranda with a friend, tucked under a huge red umbrella, gazing through dark lenses across dunes and beach to the glittering blue Atlantic. You chatter, you listen. Your lunch arrives. And a big black bird. He, or perhaps she, perches on the back of the chair directly across the table and tilts his head. "Hello," you say. You smile at the bird. You fancy that he smiles back. You and your friend watch him and laugh. He hops onto the table, tilts his head, and eyes you again. You ask, in your clever human way, "Are you hungry?"

And the bird says *Yeeees.*

His voice scrapes your eardrum, and his enunciation could use some work, but there's no mistaking the word. Just to be sure, you ask, "Would you like something to eat?" and again he says, *Yeeees.* Who could say no to that?

Still, we have to be cautious when we interpret animal behaviors, especially when we want a behavior to mean something in particular. Wanting is a drug, a hallucinogen. Even scientists trained to be cautious can be duped by their own desires. In the 1960s and '70s, much was made of attempts to teach apes to communicate with their handlers through American Sign Language. The researchers believed they had succeeded. They cited examples of clever, grammatical constructions produced by the apes, and their scientific articles were soon published in plain English to overwhelming public delight. The scientists wanted to believe, and we the public wanted to believe, but later attempts to replicate the results of those studies failed. The apes had learned something, but it was not human language.

The crow who sat down to lunch with me and my friend said *yes.* I doubt that he understood my question, or the meaning of the word he uttered, but he knew that if he made that sound we were likely to give him some food. Someone, perhaps a long line of someones, taught him the rising tones of a question, taught him to mimic human speech in response. Taught him that saying *yeees* in his gravely way might cause someone to share a bit of lunch. He delighted us enough that we repeated our behavior six, seven times, handing over bits of turkey and fruit, bread and tomato. If we looked on from another angle, we might suspect that our black-feathered friend trained *us*. As in all good training, in the end it doesn't much matter who trained whom; we all got what we wanted. Crow ate, we laughed. We tell the story for years. Perhaps Crow does too.

2. *Family Corvidae*

> Then this ebony bird beguiling my sad fancy into smiling,
> By the grave and stern decorum of the countenance it wore . . .
> —Edgar Allen Poe

Corvids, or more properly Corvidae, are a diverse bunch. Commonly known as "the crow family," they count among their number some 120 species of crows and choughs, jays and jackdaws, ravens and rooks, magpies, nutcrackers, treepies. Most corvids have voices a bit like well-amplified, well-rusted hinges, but they are still considered to be the largest of the so-called songbirds. They are more accurately "perching birds," or *passerines*. More than a third of corvid species are crows, ravens, and jackdaws, members all of the genus *Corvus*.

With brain-to-body-mass ratios that match those of the cetaceans and apes (and lag only slightly behind our own), corvids are considered by many people to be among the most intelligent animals. They certainly top the avian honor roll. Ravens and crows in particular have been seen in many traditions as divine messengers, tricksters, mediums, and omens. Tinglit, Haida, and other Native American traditions honor Raven as both Trickster and Creator. Many other traditions link corvids to war, death, and the underworld, perhaps due to their dark plumage and fondness for carrion. These associations remain alive in contemporary literature and popular culture.

As is the case with much folklore, truth lives in tales and beliefs. Ravens, crows, and magpies hunt and scavenge in life as in tradition. Hunters report that ravens and crows are quick to arrive after gunfire, adapting millennia-old symbioses between these birds and large predators, particularly wolves, to modern realities. Beyond that, corvids are messengers of death, not in some prescient otherworldly way but for practical reasons. Field studies have shown that ravens "call" wolves to large animals they find dead. Why invite wolves to dinner? Because, unlike birds of prey, the raven lacks a bill or talons designed to open a carcass. Someone else—wolf or human hunter or motor vehicle—needs to do the job. Magpies have been observed working with coyotes in much the same way as ravens work with wolves, and the canine hunters have learned to listen when corvids call.

Corvids aren't entirely dependent, though, on the kindness of other hunters. For one thing, they are omnivorous; they eat everything from insects and meat to seeds and fruits to garbage and animal feeds. Many corvids are fond (in a dietary sense) of small mammals and other birds. They prey on eggs and nestlings and are not exempt from their own cousins' raids; ravens have been seen to take apart the elaborate nests of magpies stick by carefully woven stick to get at the nestlings. We can hardly fault them (although traditionally we do); we too kill to eat. So do the raptors that commonly prey on corvids. So do many birds. Still, in Western traditions and beyond, people have both respected the cleverness of corvids and feared their presence. In Cornwall and other parts of the Celtic world, we are advised to greet any magpie we meet politely and loudly. This is, I think, good advice.

3. Corvidae Pica hudsonia *(pie-ca hud-sonya)*

> I have a magpie mind. I like anything that glitters.
> —Lord Thomson of Fleet

Something glitters in the bright spring light on the far side of Evans Creek. Something moves, and I stop. Watch. A black-billed magpie stands at the top of the far bank, wings open wide and slightly drooped, head down, tail feathers spread like a Spanish fan, back feathers raised and fluffed. Three more magpies flutter in the cottonwood to my right. I sidle into the shade of the tree. The bird on the ground shudders, steps toward a desiccated clump of rabbit brush, goes still. We think of these birds as black-and-white, but the sun on outstretched feathers reveals startling blue, at least three shades of violet, a teasing of green and orange and gold. Paiute fancy-shawl dancers come to mind, their embroidered finery outspread like wings. I watch the bird, bedazzled.

This is the first time I have seen this behavior, but I have read descriptions and know that this magpie is not dancing but "anting." Many birds do this—crows, babblers, weavers, owls, turkeys, waxbills, pheasants, more. Magpies. They pick up ants and place them on their own bodies, let them walk around or squish them like little sponges against their feathers, showering themselves in formic acid, the ant's chemical defense. The birds won't say why they do it, but scientists have several theories. The formic acid may serve as an insecticide, fending off parasitic mites, or it may help control fungal or bacterial infections. Why not? We use formic acid to fight off mites in honeybee colonies, to slow fungal growth in animal feeds, and to battle deadly *E. coli* bacteria. Other researchers suggest formic acid as grooming product or ant as vitamin supplement because it contains significant levels of vitamin D.

My bird shudders again, and I remember reading that anting appears to intoxicate some birds. Again, why not? Shamans of south-central California ingest harvester ants to induce religious visions. All functional explanations aside, perhaps some birds just like to get wasted. My magpie lifts his wings a shade, lowers them, lifts them again. He bobs his head, shakes it, stretches his neck forward, lifts his gleaming black bill skyward, says *grrrk grrrk.* As I watch him fold his wings and turn his profile to me, I don't care what moves this bird. I care only that I was witness, and that I will never again see magpies in black-and-white.

4. *Corvidae* Corvus cornix

> Light thickens, and the crow
> Makes wing to th' rooky wood.
> —Shakespeare, *Macbeth*

New Year's Eve on Balscadden Bay. It is late afternoon, and a pale sun hangs on the horizon as if hesitant to plunge into the cold Irish Sea. Gray is all around. I sit on a gray concrete wall above a beach strewn with gray rocks. A heron stands at the surf line, not the great blue or the white I know from home but *Ardea cinerea,* the gray. A flock of gulls swoops and screams over the bay; another group hunches on a trio of boulders now high above the outgoing tide.

A bird I have only just met works among the pebbles and pools, and I am enthralled. She—I don't know that the bird is female, but her industry and style make me want to think so—pulls a length of seaweed from the surf and hauls it away from the sea. This brings her closer to me, so I have a good view. She drops her treasure into a small tidal pool in the bowl of a gray hunk of rock I take to be limestone. She picks it up again a few inches from one end, works it in her bill for just the right grip. The bird swings the short end around and whacks it against the rock side of the bowl. Repeats. Again. She drops the length of seaweed and pecks around in the shallow water. I watch for 20 minutes as the gray-and-black bird repeats the process, working her way to the other end of the soggy vegetable. She flies back to the surf and I walk to her pool. The remains of the seaweed are a raggedy mess, but I poke around the strands and look into the pool, and I find what she was after. Tiny crustaceans. She hasn't left many, but a few little shells are still caught in the fibers.

This is the hooded crow, the hoodie. She goes by many other names as well, and I discover that not many people share my enchanted view of her here in Ireland. My landlady tells me they frighten her, but she can't give me a solid reason why that is so. Perhaps, she says, it is from seeing *The Birds,* Alfred Hitchcock's horror classic. Or perhaps her fear has been handed down from mother to daughter, a thread woven into the cultural fabric. For this is the corbie of ballads; a loose translation of one perches a pair of corbies on the dead hero's white breastbone to peck out his bonny blue eyes. Depending on which source we believe, the hooded crow or the raven embodied the Morrigan, Celtic goddess of war and death; when the Irish mythic hero Cú Chulainn is dying, the hooded crow perches on his shoulder. Hoodies have been variously thought to be in league with fairies, to attack livestock, and to herald death. Perhaps what puts people off is the hoodie's somber dress, black and dull brownish gray, as much as her fondness for carrion. Perhaps it is her intelligence.

The last light leaves the horizon and my hooded crow on the beach beats one last length of seaweed. I have resumed my watch from this chilly concrete perch, reluctant to leave even as evening cold seeps into my bones. Whatever legend and local opinion make of this bird, I want to know the hoodie better. The crow snatches a few last bites from the limestone bowl before her and then, with a parting *caw-caw,* she opens her wings and is gone.

5. *Corvidae* Pica hudsonia

> What I am interested in with birds . . . is what they do and why they do it.
> —David Attenborough

Bartley Ranch Regional Park lies south of Reno in the shadow of Mount Rose and her sister peaks of the eastern Sierras. On a good day the park's population includes walkers and dogs, trail runners, horses and riders, jackrabbits and cottontails. Coyotes, although they are wary. In the year I have lived here, I have only ever seen one coyote in the park, and that one a long way off. There are reptiles too. Horned lizards, and snakes—gopher snakes mostly, despite widespread fear of rattlers. Ground squirrels, marmots, other furry things. Birds.

Golden eagles ride the thermals over Bartley. Vultures too. Red-tailed and Cooper's hawks are common, and owls—great horned and burrowing and barred. California quail are everywhere, strutting and chattering in the brush, or hollering *ne-VA-da ne-VA-da* at everyone, no matter the birds' common name. Crows and songbirds of all sorts. Magpies.

At first I think the bird is injured. I am walking along the edge of a parking lot, headed for the trail that climbs up and back down Windy Hill, when a wild fluttering catches my eye. A wingtip and tail's long feathers wave and shake at me from behind a hassock-sized boulder. Magpie, I think, and a vision of the anting magpie dances across my memory. Magpie in trouble. I will check the bird and, if necessary, call for a ranger. Then a scream from behind the rock clutches at me. Not a magpie scream. Before I see, I know.

The magpie is not in trouble, and even as I rush toward them, she pecks at the little rabbit. Another scream. New from the nest, smaller than my fist, covered with downy gray-brown fur and, now, a spattering of blood. *Get off,* I yell, waving my arms at the bird, feeling a surge of adrenaline and something else, something that makes me want to hold that baby, save him, stop the hurt and fear. The magpie slowly lifts her body into the air and sinks like a hovercraft onto a branch just over my head. I could reach out and touch the long tail from where I stand. A hard black eye bores into me. As I turn back to the baby rabbit, the magpie rasps *mag-*

mag-mag-mag, and I wonder vaguely whether she might not come for me. The bunny is still, but its eyes are alive. I shoo it from the open space behind the boulder into a thick tangle of sage. It moves quickly and disappears. Safe for the moment, I think.

The bird on the branch is still glaring when I look up again. I am no coward about the facts of nature. This is what animals do, all of us, and even as I wave my arms to drive the bird away, knowing she will be back, I ponder my right to do such a thing. I think about such things a lot when I walk. Part of the path I meant to take this morning lies between two pastures where feeder calves graze the summer away, and because I walk there, and because they are no longer visions in a distant field but individuals who come to the fence to watch *me*, whose faces I know—the red steer with the question mark on his brow, the black one with the punky topknot that stands between his ears—I have been grappling with my own conflicted habits.

My heart slows and I turn away from the sagebrush, away from the birdless branch. I halt. The sand and high-desert plants around me are in wild motion. Magpies everywhere, 30 or more, swooping, pecking, chasing baby rabbits. My ears fill with the screaming of the bunnies, the raucous whoops and caws of the birds. Two adult rabbits—the mothers, I know in my heart—run in wild hopeless loops across the open spaces. I begin to run too, yelling at birds and waving my arms and stopping to herd the terrified babies into sage and rabbitbrush and spaces beneath rocks. I know it will all start again when I leave. But in this moment, it is what I do.

6. *Corvidae* Pica hudsonia

> . . . the great and flashing magpie, He flies as poets might.
> —T. P. Cameron Wilson

Lewis and Clark met their first magpies in South Dakota in 1804. They wrote that the birds were bold and gregarious, willing to walk right into the men's tents and take food from their hands. The collected specimens they sent east to Thomas Jefferson included four live magpies, although only one made it to Monticello. When bison roamed the Great Plains, magpies went along, picking ticks off the great beasts and eating insects the herds stirred up. The birds scavenged as well, cleaning up carcasses left by people and wolves and other death-dealers. The bald eagle may be the avian symbol of the West, but it is the black-billed magpie who will dance for you on creek banks and call *mag mag mag* as you walk the riparian strands.

Smart and adaptable, magpies switched to other livestock when men with rifles wiped out the great herds of bison in the 1870s. They are supreme opportunists. I have watched flocks of magpies hunting on creek banks and picking insects and bits of grain from manure in horse paddocks. Adaptability to human-dominated environments has advantages. It also brings new dangers. Campaigns to eradicate these "pests" have caused thousands of magpie deaths; 1,033 black-billed magpies were shot in the Okanogan Valley of Washington in one "hunt" in 1933, and in Idaho an estimated 150,000 magpies were systematically killed around the same time for a few pennies in bounty per bird. Thousands more have died from eating poisons set out for coyotes and other predators.

A modified war on magpies continues today. As is the case in wars, hatreds often rest on false beliefs. Magpies don't peck livestock open for the blood; they pull off and devour ticks. Magpies don't decimate "desirable" songbird populations; they all thrive together. Facts take a long time, though, to overshadow falsehoods, and although magpies in the United States are partially protected under the Migratory Bird Treaty Act and some state and local statutes, judgment-call exceptions are written into the law.

7. *Corvidae* Corvus brachyrhynchos

> I'd hate to miss an important message
> because it came dressed in feathered black.
> —Lynn Samsel

The new issue of *National Wildlife* has arrived, and in it is a story ostensibly about crows. Really, though, it is a story about people, or, more to the point, scientists, and about their "discovery" that crows recognize human faces. Not as a class of things, mind you, but individual faces. That crow who seems to know you, well, she knows *you*. This, apparently, is a surprise. At least that's the impression the researchers give in their comments, which seem crafted to fit a rhetoric of objective distancing. Their surprise at their findings implies instead a disconnection with the creatures being studied. Perhaps the question should not be *whether* the crows are able to recognize individual faces, but *why*.

Scientists are wary of anthropomorphizing. We all should be. But sometimes researchers seem to be even more wary of being accused of the act than they are of the act itself. It gets plain silly at times. Eight pages after the article about crows is another that asks, "Are Other Animals Aware of Death?" Have you ever watched an encounter between a predator and its prey? Or seen a mother animal with a dead or dying baby? I have, and would say that they know at least as much about death as we do.

Elephants are the focus of the piece. They often are when nonhuman awareness of death comes up, because they are known to linger over and handle the bones of their own dead. Corvids are mentioned as well. A recent study at the University of California, Davis reported that scrub jays (*Aphelocoma californica*) will gather around the remains of other scrub jays and scream for up to 30 minutes. Initially the ruckus was thought to be a warning, but jays, and sometimes ravens and crows, respond to the uproar by flocking *to* the sites. Still, so goes the report, the gatherings "don't necessarily mean the birds understand death." Whether they understand death seems to me as silly as asking a crow at table "Are you hungry?" After all, our own species has been debating what death is, biologically and philosophically, for thousands of years. How far ahead of the jays does that put us in coming to terms with mortality?

"Would you like something to eat?" I asked the crow at the beach. Why else would he bother with me? I need to reform my questions. The sharp-eyed corvids may be clever enough to help if I can manage to listen beyond what I want to hear. Or perhaps they know that some answers are entirely avian and beyond our reach.

REBECCA BOYLE

The Health Effects of a World Without Darkness

FROM *Aeon*

SOUND DOMINATED MY senses as we left the village of San Pedro de Atacama and walked into the desert night. The crunch of shoes on gravel underlaid our voices, which were hushed to avoid waking any households or street dogs. Our small group of astronomy writers was escaping from light, and without any flashlights or streetlamps, we struggled to see, so our other senses were heightened. Land that looked red by day was now monochromatic, the rods in our retinas serving as our only visual input.

After about 15 minutes of hiking, we stopped to take some pictures of the sky. I fumbled with my gear and tried to get my bearings, but everything was alien. I was horribly jetlagged after 10 hours hunched against the window of a 757, another two-hour flight north from Santiago, and a two-hour bus ride, and it wasn't just my oxygen-hungry brain that put me out of sorts. The Atacama Desert looked like Mars as drawn by Dr. Seuss; I was surrounded by wrong-colored cliffs and swirling rock formations. But I was determined to photograph something even more bizarre: the Large Magellanic Cloud, a dwarf galaxy you can see only from the Southern Hemisphere. I perched my camera on a rock and aimed at the sky, but the cosmic smudge would not resolve in my viewfinder. I stood, brushed dirt from my jeans, and looked up.

The unfamiliar sky momentarily took away what little breath I had left at 8,000 feet in elevation. Above the horizon was the conspicuous Southern Cross. Orion was there too, but looked as disoriented as I felt, upside down to the world. And there were so many constellations I'd never seen, with hopeful, Latinate names such as Dorado and Reticulum. Countless stars blazed into view as I stared into the smear of the Milky Way.

To most people who have traveled outside the developed world —whether to camp or to meditate or to hunt—such bright and plentiful stars are a glorious sight. But this beauty instilled in me a creeping sense of guilt. At home, 1,500 miles north, I wouldn't recognize such spangled heavens. From where I live in the American Midwest, the stars might as well not exist. After journeying millions of years, their light is swallowed by city glare and my porch lantern. Those that make it through will still fail: not even bright Betelgeuse can outshine my iPhone. Yet I am an astronomy writer, a person who thinks about stars and planets all the time. What does my neglect of the night sky say about the rest of humanity?

"We are all descended from astronomers," the astrophysicist Neil deGrasse Tyson intones in the rebooted version of the TV show *Cosmos*. This is as poetic as it is true. Everyone owns the night sky; it was the one natural realm all our ancestors could see and know intimately. No river, no grand mountain or canyon, not even the oceans can claim that. But since Edison's light bulbs colonized our cities, the vast majority of humans have ceased to see those skies. More than 60 percent of the world, and fully 99 percent of the U.S. and Europe, live under a yellowy sky polluted with light. For many of us the only place to see the milky backbone of our own galaxy is on the ceiling of a planetarium. Although humans are diurnal, factories and Twitter and hospitals and CNN are not, so we must conquer the darkness. As a result, almost everything industrialized people build is lit up at night. Malls, hospitals, car dealerships. Streets, bridges, air and sea ports. Buildings on a skyline. These artificial lights identify our cities all the way from the moon. If aliens ever do drop by, this might be their first sign that someone is home.

But cosmology, the study and interpretation of the universe, has always depended on a star-choked dark sky. Ancient civilizations from the Greeks to the Pawnee looked to the stars and saw not only creation tales but active participants in their lives. Christians, who invest great meaning in the good of light and the evil of darkness, spread a starry message too: the star of Bethlehem as a beacon to salvation. A millennium and a half later, Galileo looked up and saw a new version of the cosmos, breaking the dawn of modern science. And Edwin Hubble discovered the expansion of the universe by the candlelight of supernovas. All of this happened under virginal skies, and by any measure we don't have those anymore. We look at our glowing rectangles, and we opt out of that shared heritage.

Nowhere is light pollution more apparent, almost achingly so, than in satellite images of Earth from space. The continental United States seems to split in half: the eastern side is brighter than the West, except for the klieg lights of Las Vegas. Highways innervate America, connecting luminous dots of small towns and big cities. Across the Atlantic, Europe shimmers. Moscow is a radiant nine-pointed star. The Nile Delta glows like a dandelion sprouting from mostly indigo Africa. Farther east, Hong Kong and Shanghai are ablaze, and the demilitarized zone separates dark North Korea from South Korea more cleanly than if the peninsula had been cleft in two. Developed society, it's clear, is where the light is.

Human-controlled light has pierced the night for thousands of years, long before Edison. Campfires warmed our ancestors' feet and cooked their meals; the Harvard anthropologist Richard Wrangham argues in his book *Catching Fire* (2009) that gathering around a flame to eat and to commune with others is in fact what made us human. Not just fellowship but safety has long been the primary rationale for pushing back the night. "Evil spirits love not the smell of lamps," as Plato put it. Comforting, lambent lamplight led us safely home by tattling on the people and potholes and animals that would otherwise do us harm. By the early 17th century, residents of cities such as Paris and London were admonished to keep lights burning in the windows of all houses that faced the streets, as the historian A. Roger Ekirch notes in his book *At Day's Close: Night in Times Past* (2005). Taxpayers funded oil lamps and candlelit lanterns for the avenues, while only genteel households could afford fine beeswax or spermaceti candles; most people relied on tallow, made from animal fat.

Despite their utility, these artificial lights were sources of danger in their own right. Huge swaths of cities—notably London and Chicago—were consumed in conflagrations that started as accidents, born of the necessity of using flames to see. By the 1800s gas lamps reduced fire risks, but cities were by no means safer from crime; gas-lit London in the late 1880s, full of foggy halos casting shadows down dark alleys, is as famous for murder as anything else. Even now, artificial light provides an artificial sense of security. A 1997 report from the U.S. National Institute of Justice found no conclusive correlation between nighttime lighting and crime rates. The International Dark Sky Association, a dedicated group of nighttime advocates, points out that bright, glaring lamps create sharper contrasts between light and darkness, blinding drivers and homeowners alike.

And even so—what price safety! A young but rapidly growing field of research suggests that nighttime light itself is far more dangerous than the dark. In a 2012 report an American Medical Association committee called electric lighting a "man-made self-experiment" creating potentially harmful health effects. Humans, and everything else that lives on this planet with us, evolved during billions of years along a reliable cycle of day and night, with clear boundaries between them. Stanching the flood of artificial light can help restore this divide. Our well-being, and that of our fellow creatures, might depend on us doing so—or at the very least trying. The loss of nighttime darkness neglects our shared past, but it might very well cut short our futures too.

The midnight desert was quiet while I knelt with my camera last spring, but the Atacama was far from asleep. Beetles and red scorpions scuttled across the dirt. Vallenar toads crouched on the *lomas.* South American gray foxes sniffed the earth, hunting furry viscachas and Darwin's leaf-eared mice. Great horned owls circled overhead, hunting the rodents and the foxes. Nocturnal animals such as these make up 30 percent of all vertebrates and 60 percent of all invertebrates on Earth, according to an estimate by the German biologist Franz Hölker and colleagues. These night-dwellers are the most obvious victims of artificial light. Light pollution interferes with their natural rhythms in myriad ways.

To gauge levels of light pollution, scientists use lux, which is a measurement of illuminance that counts how many photons per second strike our eyes. As an example, the planet Venus, at its brightest, produces 0.0001 lux. In the natural nightscape, plants and animals are exposed to light levels that max out around 0.1 to 0.3 lux, during the week around full moon. By contrast, a typical shopping mall gushes 10 to 20 lux at night. That is 200,000 times brighter than the illuminance of a moonless evening.

For migratory birds that fly at night, artificial light is a deadly siren. In New York every September, columns of light shine skyward in tribute to the destroyed World Trade Center towers. Tens of thousands of migrating birds, trying to navigate by the moon and stars, fly into the beams and circle, zombielike, until someone shuts the lights off. Birds also collide with glittering buildings and lighthouses and are stunned, falling to their deaths. It is such a widespread problem that cities from Toronto to Chicago adopt lights-out campaigns during peak migrations.

Sea turtles also need a dark sky atlas to find their way. Newly hatched on the Atlantic coast, they are confused by beaches bathed in light and follow a false moon, turning away from the safety of the sea. Florida wildlife officials and even NASA have spent decades trying to build better beach embankments, using old railcars, driftwood, and sand dunes to mask the artificial light streaming from highways and launch pads.

Almost all bat species are nocturnal, hunting out insects, frogs, nectar, pollen, fruit, and other bats when it is dark. Omnivorous bats use echolocation rather than vision to track their prey, but extra light is far from helpful. Insect-eating bats chose different foraging routes to avoid just 0.4 lux of light, according to a 2009 study led by Emma Stone of the University of Bristol. Fruit bats avoid the glare too. Costa Rican short-tailed bats, given a choice between pepper plants growing in the dark and plants illuminated by sodium lamps, chose the dark twice as often. There are ecological consequences: changing their flight paths alters the "seed rain," showered by defecating bats, that can be crucial to recolonizing clear-cut rainforest. Along with changing the eating habits of bats, light can have direct physiological effects, another study found. Juvenile bats that hailed from illuminated buildings were smaller—their wings were shorter and they weighed less—than those born to the dark.

For pelagic ocean animals, who live in the liminal space between the surface and the sea floor, the light of the moon and the sun are the only landmarks. Dazzling boats in coastal waters—another luminous activity visible from space—lure fish to the surface and into nets but also interfere with marine organisms' navigation, hunting, and mating habits.

Including moths to a flame, nocturnal invertebrates are the

most well recognized examples of creatures disoriented by light. Insects congregate around light sources until they die of exhaustion (or, caught in the spotlight, are eaten by birds and bats). In a 2012 study the Exeter scientist Jonathan Bennie found that light pollution changed the composition of ecological communities among five major invertebrate groups. "Street lighting changes the environment at higher levels of biological organisation than previously recognised," Bennie and his coauthors wrote, "raising the potential that it can alter the structure and function of ecosystems."

This is true for diurnal creatures too. Animals that make their living during the day are still disrupted by artificial nighttime light. Occasionally artificial light can be beneficial by extending the hours of day—in a 2012 study the biologist Ross Dwyer and his colleagues at the University of Exeter found a waterbird called the common redshank foraged longer and more effectively at night along an industrialized Scottish estuary. But as with the insects, this carries consequences for the broader ecosystem.

On land, artificial lighting causes a cascade of negative physiological changes in diurnal creatures, many brought about by delayed release of the hormone melatonin. Davide Dominoni and his colleagues at the Max Planck Institute in Germany found it was suppressed in European blackbirds exposed to just 0.3 lux at night. These birds developed their reproductive systems a month earlier, and molted earlier, than birds kept in the dark. A different study, by Travis Longcore of the Urban Wildlands Group in Los Angeles, found that blue tits under the influence of streetlamps laid their eggs earlier than those experiencing dark nights. And in mammals, from mice to men, the effects of melatonin suppression might be far worse.

A growing body of evidence shows that light pollution exacerbates, and might directly cause, cancer, obesity, and depression, the troublesome triumvirate of industrialized society. One of the first people to notice this correlation, at least as it applies to cancer, is Richard Stevens, a professor at the University of Connecticut, respected cancer epidemiologist, and mild insomniac. In the early 1980s Stevens and other researchers were beginning to realize that there was little or no connection between diet and rising rates of breast cancer, contrary to what had been suspected. As

Stevens puts it, it was like a light bulb going on when he realized that in fact a light bulb going on might be a culprit.

"I really did wake up in the middle of the night in my apartment in Washington State and realize I could read a newspaper by the light from a streetlight. And I wondered, what's that?" he told me. "So I started calling around. I started to learn about circadian rhythmicity. And about this hormone called melatonin." His 1987 paper "Electric Power Use and Breast Cancer: A Hypothesis" was one of the first to report the potential connection between rising cancer rates and artificial nighttime light exposure, something he and others have continued to report in the intervening 27 years.

Melatonin is produced in the pineal gland, a small pine-cone-shaped knob in the center of the vertebrate brain. It is derived from serotonin, a neurotransmitter that is involved in mood and appetite. And melatonin is an antioxidant, which protects DNA from damage; this has important implications for cancer biology. Stevens has published research demonstrating that melatonin can prevent breast tumors in rats. But the hormone's chief role is in regulating the daily sleep-wake cycle by causing drowsiness and lowering core body temperature. Melatonin has the same basic function in people, birds, fish, amphibians, and other mammals. Production of melatonin should begin at dusk, when we are supposed to sleep. Light—not wakefulness itself, but light—shuts it off, as Stevens emphasized to me.

A remarkably recent discovery helps explain what's going on. In 2000 scientists noticed a light-capturing pigment, which they called melanopsin, in the retina. It's a different pigment from the types in our cone- and rod-shaped photoreceptors, which helped me see the monochrome landscape of the midnight Atacama. In 2002 the biologists David Berson of Brown University and Samer Hattar of Johns Hopkins University rediscovered a special type of cell that uses this pigment (they were first described, but then forgotten, in 1923). They were dubbed "intrinsically photosensitive retinal ganglion cells"—simply put, they have nothing to do with vision, but they sense light, whose presence they communicate directly to the brain. Blue light, to be specific. By studying the eyes of primitive creatures such as lampreys and hagfish, scientists can tell that these special cells have been present in the vertebrate retina for at least 500 million years. This means that since the dawn of backboned animals, all such creatures have been equipped to track the day and night and calibrate their metabolic cycles accordingly.

When you think about the fact that humans figured out fire 250,000 years ago and electricity just 130 years ago, the importance of light to our brains and biology starts to become clearer. As Stevens puts it, circadian biology is at the core of all biology, human biology included. Randy Nelson, a circadian biologist at Ohio State University, has been studying light's effects on depression and obesity since 2004, when one of his graduate students was hospitalized for a staph infection. The student complained bitterly about the bright lights in his room and in the hospital hallway, which robbed him of sleep and stressed him out. Nelson and another graduate student, Laura Fonken, decided to investigate this complaint using rodents as experimental subjects. They found that mice who were exposed to constant bright light exhibited depressive symptoms, behaving listlessly and ignoring their sugar-water treats. Remarkably, they then found that the same happened when the mice were exposed to only 5 lux at night, when the animals were normally active. This is equivalent, Fonken notes, to leaving a television on in your bedroom or a computer screen next to your head as you nod off. Later the team worked with diurnal Nile grass rats instead of nocturnal mice and found the same thing. The rats exhibited not only depression but demonstrable changes in neuronal connectivity in the hippocampus, the part of the brain involved in learning, memory, and affective responses.

Along the way Fonken also noticed something unexpected: the light-exposed rodents got fat, even though they were eating the same number of calories as their dark-sequestered mates. What changed was their circadian rhythms; like a snacky night owl, they were eating when they should have been inactive, upending their digestive and metabolic activity. One side effect of inbreeding lab mice is that some of them do not produce melatonin, Fonken told me, which means something else might have been interrupting their internal timepieces. Fonken looked at gene expression and noticed changes in a gene known helpfully as CLOCK (Circadian Locomotor Output Cycles Kaput), among others. Messenger RNA molecules were activated by light, switching multiple genes on or off to regulate circadian rhythms. Fonken guards against drawing too many human conclusions from these rodent studies, but epidemiological evidence suggests the implications for people could be profound.

"Rather than falling, night, to the watchful eye, rises," writes Ekirch in *At Day's Close.* Shadows creep up lows and valleys first, then consume hillsides and houses and the tallest buildings. Muted grays and deeper blues chase off the sun until finally the sky leaks no color. When we sleep according to a solar cycle, melatonin production follows this pattern, rising with the night. But artificial light tamps it down. This is frustratingly apparent for a special class of humans who experience sunsets every 90 minutes: astronauts.

One of the most frequent complaints of orbital crew members is insomnia; they pop sleeping pills on a regular basis and still get only about six hours of shuteye, though they're allotted eight. Steven W. Lockley, a Harvard neuroscientist, recommends altering the light to improve matters. In 2012 he advised NASA engineers to change the light bulbs on the International Space Station to a type of LED that can display blueshifted light during the "day," when the crew is working, and redshifted light when they need to rest. Why the difference? That crucial ganglion, the circadian photoreceptor, is particularly sensitive to light toward the bluish end of the red, orange, yellow, green, blue, indigo, and violet visible-light spectrum.

Blue light also pours from the phone, tablet, or computer screen on which you're reading this. This light, in a wavelength very similar to daylight, has been shown to exacerbate insomnia in scores of studies. In 2007 Belgian researchers surveyed 1,656 teenagers about their use of mobile phones after lights-out and found that those who used a phone *less than once per week* were more than twice as likely to be "very tired" a year later as those who never did. Using a phone after lights-out about once a week increased the risk of being "very tired" by five times. In 2012 researchers at Rensselaer Polytechnic Institute showed that two hours of exposure to a bright tablet screen at night, like an iPad or a Kindle, reduced melatonin levels by 22 percent. And the National Sleep Foundation says that more than 90 percent of Americans regularly use some type of electronic device in the hour before bed.

There is a great difference between natural night waking and electronic-induced insomnia, Stevens points out. In his history of the night, Ekirch explains that temporary nighttime waking—of the type that used to plague Stevens—is hardly unnatural. Humans have aroused in the middle of the night since time immemorial to pee, snack, chat with relatives and neighbors, make love, and simply enjoy "quiet wakefulness." "There is every reason to believe that segmented sleep, such as many wild animals exhibit, had long been the natural pattern of our slumber before the modern age, with a provenance as old as humankind," Ekirch writes. The unnatural part is that now we get up and we turn on the light, silencing melatonin.

Insomnia is hardly the worst side effect of light pollution. Shift workers, who rise with the night and work awash in blue light, experience not only disrupted circadian rhythms and sleep deprivation but an increased risk of breast and prostate cancer. These cancers, which require hormones to grow, are suppressed in the presence of melatonin, Stevens has shown. In 2010 Stevens published a review of breast cancer sensitivity in 164 countries and found a 30 to 50 percent increased risk of cancer in nations with the worst light pollution, but no increased prevalence of non-hormonally dependent cancers in the same populations.

"Our use of electric light in the modern world is disrupting our circadian sleep and our biology. There is no question about that. Does that have physiological consequences? There is more and more evidence that it does," Stevens told me. "The epidemiological studies are the crudest but the most important."

As we discussed a litany of light-related problems, I asked Stevens, "Is it a legitimate question to ask if light is the major factor in depression, obesity, and cancer? Is there potential for light to be *the* reason behind all of those things?"

"Yes," he said flatly. "No doubt about it." The day is coming when doctors might feel confident saying so, he added, just as they now say that smoking causes lung cancer. The murky part is what to do about it.

A torrent of light is what you might call a "First World problem." Even as Western cities try to bring back starlit nights, students throughout developing countries in Southeast Asia and sub-Saharan Africa still read by candlelight. Economically depressed regions tend to be darker; after the collapse of communism, satellite imagery showed several countries in the former Soviet Union were dimmer at night.

When we in the industrialized world do manage to turn off the lights, there are measurable, beneficial effects on our circadian rhythms. In a widely reported paper last summer, Kenneth Wright at the University of Colorado at Boulder took eight students camping in the Rocky Mountains for two weeks. They weren't allowed to use any artificial light after the sun went down—only the sanguine glow of campfire. After a week melatonin started to rise at sunset, peak in the middle of the night, and taper just at sunrise, which Wright called a "remarkable" result. "Internal biological time under natural light-dark conditions tightly synchronizes to environmental time, and in this regard, humans are comparable to other animals," he wrote.

Of course, very few of us are going to camp outside nightly or stop working at 4:30 p.m. in winter, let alone shut down power plants and global commerce. And even as I mourn the loss of night, I am no Luddite myself. I read Wright's study and so many others while sitting in bed, hours after the sun had set, bathed in the bluish spectrum of light emitted by my iPad. I am a nighttime chronotype and I drink electricity as thirstily as anyone.

But there are a few things we can all do.

As NASA has found, redshifted lamps, rather than a blue hue, are a better choice for bedside reading. Chronobiologists are concerned about the fluorescent swirls that will comprise the majority of American light fixtures when the ban on incandescent bulbs finally sets in. They emit too much blue light—we should use redshifted LEDs instead. The same goes for many building lights and streetlamps. Along with keeping us more awake, this light is especially intrusive for the same reason the sky is blue: it scatters more readily, causing even more light pollution.

Downward-directed streetlamps can illuminate larger areas while casting fewer shadows and errant upward glow. Smarter, more efficient use can cut wasted light, wasted money, and wasted energy. This has been a powerful factor in bringing back the night in towns from Brainerd, Minnesota, to Paris, the City of Light. Since July 2013 the French capital has been going dark for several hours at night to save money and "reduce the print of artificial lighting on the nocturnal environment," in the words of its Environment Ministry. In a paper published this January, Exeter's Jonathan Bennie and colleagues reported that some economically developed regions, especially throughout Europe, are trending darker.

For those of us addicted to our glowing phones and tablets, an app called F.lux can help. It "warms" your device's display screen so that it shifts red in the evening, more closely matching incandescent bulbs and the hue of the setting sun. Just a brief glimpse at your mobile phone at bedtime is enough to expose your retinas to artificial light, so fighting such a comprehensive intrusion might be an exercise in futility. But even if we can't completely quench our thirst for light, we can all make one small gesture, which could prompt us to unplug a little more.

I try to do this whenever the sky above St. Louis is clear. After I climb out of my car but before I open the fence gate and the hinge cheeps, before I step onto the deck and the wood sags and the motion-sensor light turns on, I stop, and I look up. I nod at Jupiter, blazing in the east. I greet Orion in his more familiar northern form. I squint at the moon if it is there and I let my eyes open wide, hoping more stars will leak through the city haze and reach me standing there, in the darkness.

ALISON HAWTHORNE DEMING

Spotted Hyena

FROM *Orion*

ON A DARK and moonless night of storms, Hans Kruuk observed a pack of spotted hyenas engaging in a predatory orgy of killing. The Dutch biologist had spent three and a half years living with Serengeti hyenas, documenting their complex matriarchal clan structure and cooperative hunting strategies and keeping a full-grown hyena named Solomon as a house pet. Hyenas make a living by scavenging and hunting. Their large jaw muscles and bone-crushing premolars give them chewing ability stronger than a brown bear's. They eat and digest parts of dead animals other predators leave for waste, including bones. Their gut is so efficient that their feces are white and powdery. Hyenas rival primates in intelligence and skill at cooperative hunting. In captivity, pairs of hyenas, given the challenge to pull two ropes in unison to get a food reward, learned the skill more quickly than chimps given the same task. They rivaled the chimps too in passing on the skill to inexperienced clan mates.

Spotted hyenas are weird-looking. Cute teddy-bear ears. Fluffy, mottled coat. The cantilevered stance, hind legs much shorter than forelegs, makes the animal look somehow taller than itself and always ready to lunge. There is menace in the darting eyes. Their whooping cry has earned them the nickname "laughing hyena," though it sounds nothing like laughter except in bedlam. They live in clans of five to 90 members, led by an alpha female whose offspring will inherit her role. Aristotle thought hyenas resembled hermaphrodites. There is little sexual dimorphism in their genitalia. The female has a six-inch clitoris that can become erect. The labia are fused in a sac resembling a scrotum. This confusing equipment entered oral tradition as protoscience: in medieval bestiaries, the hyena was said to change its gender at will or with the season. For the female hyena, this gear means difficult childbirth, as the young emerge through the clitoris.

Kruuk observed that during a violent storm in Tanzania's Ngorongoro Crater, a herd of Thompson gazelles became frenzied and disoriented. The hyenas became excited and gave chase, pursuing the gazelles with a zeal matching the storm's intensity. They slaughtered 80 in a night—far more than the pack could possibly eat. He named the behavior "surplus killing," and it is now known that many carnivores do this. It may be more likely to occur when predators are overstimulated and prey weakened by ill health or weather. Kruuk later reported that three crows went on a binge, caching 79 mice in slightly over two hours. Other reports of surplus killing tally red foxes in north Scotland killing 200 black-headed gulls in a night, and similar binges by minks feeding on terns, killer whales on seal pups, and even among spiders hunting flies. Surplus killing is not the norm, but it is common enough in the carnivorous world to say that the capacity for this behavior is a component of the animal spirit.

Surplus killing looks brutal and gratuitously violent to the human eye, though our species too is quite accomplished in acts that have earned those descriptors. But for an animal that must kill to live, it makes sense for the hunt and the kill to be pleasurable. If you don't kill, you don't eat, and if you don't eat, you die. Nature's clever trick is to make behaviors that enhance survival among a creature's keenest pleasures: eating, having sex, maternal bonding, social cohesion. Animals engaged in surplus killing may simply be thrilling in their physical being—their skill and strength and muscular joy. Human beings add a new element to the repertoire of the animal spirit. In finding the behavior excessive, vicious, causing unnecessary suffering and death, humanity identifies itself as a creature holding values and making ethical judgments. Just because it feels good doesn't mean you should do it.

I visited the Ngorongoro Crater in the company of two Peace Corps volunteers, a traveling companion, and a Chaga guide. My safari offered no opportunity for the sustained observation I so admire in a researcher of Kruuk's dedication. What conclusions could I draw from a week of casual observation made through the roof of a Land Rover?

Surely not science. A casual observer can testify only to the moment. And what one sees will always be colored by what one longs to see. The unsettling truth is that most visitors to the last great animal spectacles on Earth long to see the kill. It thrills and arouses the imagination. It stimulates an atavistic hunger to live in a body so perfectly suited to meeting its needs. I saw no kill and counseled myself that I should not be disappointed.

Yes, I saw the dried pelt of a gazelle hanging from a baobab tree where a leopard had left it after supper. Yes, I saw the dismembered head of a zebra lying in scrub, the red gash where it had been severed as loud in mind today as when I saw it through the window of our racing microbus. But what struck me most keenly was the peace of the animals. Cloudlike herds of zebra, impala, and wildebeest drifted across Ngorongoro's great grassy mind, shape-shifting throughout the afternoon. Two cheetahs trotted in purposeful single file toward the opening in the crater that leads to the Serengeti. The pride of lionesses lay on their backs in the high heat of the day, forelegs dawdling in the air. The black rhino slept, imitating a boulder in tall grass. The Egyptian goose and crown-crested cranes, Fischer's lovebirds and superb starlings, hammerkops and saddle-billed storks dabbled in the shallows of wetlands or clawed up playful clouds of dust. The troop of baboons sat quietly under a sprawling tree, waiting out a rain squall. The peace of the land, the last islands of this peace, made me feel small. I welcomed the feeling. It was a pleasure to feel insignificant, to let my desires quiet, to feel, in the moment, the human body as an instrument attuned to peace.

SHERI FINK

Life, Death, and Grim Routine Fill the Day at a Liberian Ebola Clinic

FROM *The New York Times*

SUAKOKO, LIBERIA — The dirt road winds and dips, passes through a rubber plantation, and arrives up a hill, near the grounds of an old leper colony. The latest scourge, Ebola, is under assault here in a cluster of cobalt-blue buildings operated by an American charity, International Medical Corps. In the newly opened treatment center, Liberian workers and volunteers from abroad identify who is infected, save those they can, and try to halt the virus's spread.

It is a place both ordinary and otherworldly. Young men who feel well enough run laps around the ward; acrid smoke wafts from a medical waste incinerator into the expansive tropical sky; doctors are unrecognizable in yellow protective suits; patients who may not have Ebola listen to a radio with those who do, separated by a fence and fresh air.

Here are the rhythms of a single day:

7:20 a.m.

Soon after their arrival, about a half-dozen doctors and nurses gathered near whiteboards for the handoff from the night shift. There were 22 patients, and no deaths overnight. The center

—which includes a triage area, a restricted unit for patients suspected of having Ebola infections, and another for those in the grip of the disease—is not teeming like some clinics in Monrovia, more than four hours west. It is designed to accommodate up to 70 patients, but it is still scaling up after opening a few weeks ago and has just two ambulances to ferry patients.

An eight-year-old boy had been too weak to lift a liter bottle of oral rehydration solution to his mouth through the night. Bridget Anne Mulrooney, an American nurse, reported that she gave him a smaller bottle and sheets to keep warm. A woman who had lost both her baby and her husband to Ebola and was suspected of having the disease herself was refusing food and medications for symptoms and other possible illnesses, such as malaria. A man in his 70s, a talkative staff favorite, was now confused, his sheet covered in blood. He had been admitted four days earlier, but laboratory tests confirming an Ebola diagnosis had not come back yet. "I think he's positive," said Dr. Colin Bucks, an American. "I think this will be an end-of-life event."

Eight patients needed intravenous fluids to combat dehydration. One patient was described as happy. Another was playing cards.

7:40 a.m.

Morning devotion began with a song and clapping, performed in triple time. About 18 local workers, most wearing rubber boots and blue hospital scrubs bleached so often that they were now pastel pink, danced and then prayed for God's mercy on the treatment unit and those who worked there. Some folded their hands, sheathed in bright-colored gloves, at their heart. In unison the Liberians sang, "Cover with your protective arms, O God."

8:10 a.m.

Sean Casey, an American who is the center's team leader, gathered his department heads for what became a conversation about patient flow. The head of the ambulance crew said five patients with possible Ebola infections were awaiting rides to the center.

But the ward with suspected cases was full, Mr. Casey said, and needed to be cleared first. Lab results were required, so patients without Ebola could be discharged and the confirmed cases could be moved to the other ward. The center also had some patients who were ill with other maladies. They should have been transferred to the local hospital, but it offered only limited care since reopening after six nurses died of Ebola.

The managers also discussed labor issues involving the 175-member Liberian staff, some of whom had walked out days before in one of the pay disputes common among the country's health workers. The leaders and those who abandoned patient care would not be rehired, because it was crucial to have a dependable staff, Mr. Casey said. Then he sent everyone off.

"Go forward and do well," Mr. Casey instructed.

8:40 a.m.

A Liberian woman scooped steaming yam porridge out of a blue bucket—breakfast for the patients and staff. The food is prepared off-site, at a university that is closed because of the outbreak and houses many of the staff members. The center has people working as cleaners, sprayers, and waste removers—part of the so-called WASH (water, sanitation, and hygiene) team—who continuously disinfect the site and remove contaminated material. Still, the sight was a little jarring: the woman was putting the food into plastic foam plates just a few steps from the dressing rooms for staff members coming out of the decontamination areas, the pharmacy, and past a refrigerator with a sign marked EBOLA BLOOD TESTS. NO FOOD.

8:45 a.m.

The medical staff—including an American doctor, a Spanish nurse, and a Kenyan nurse, along with a Liberian physician assistant, nurses, and other workers—put on protective equipment piece by piece to enter the treatment areas. Gloves, Tychem suits, masks, hoods, aprons, goggles, the ensemble checked in a mirror to ensure no skin was showing. The process took about 20 minutes.

A doctor, Steven Hatch, entered the ward of patients with suspected Ebola infections and asked them if they were hungry. "How do you feel?" he said to one. A woman in a T-shirt walked stiffly behind him carrying a plastic chair, which looked heavy in her weak hands. Staff members outside the patient area dropped in bottles of water and pitched plastic garbage bags over a gate from the gowning room, never touching. Cleaners had gone in ahead of the medical team and were spraying the ground with chlorine solution and picking up garbage with buckets that looked just like the ones that held the yam cereal.

Staff members observing outside debated whether a set of wooden game tables with dice could be given to patients in the suspect ward. Mr. Casey decided against it, for fear of sickening someone not infected; only those confirmed Ebola patients could safely play together. Friederike Feuchte, a German psychologist, understood but was disappointed. "They feel bored," she said of the patients awaiting their test results.

One of them, Kolast Davies, 45, agreed. "Being here is stressful and very boring, especially when you don't know your fate," he said in an interview. He was admitted after managers of the steel production company where he works, ArcelorMittal, sent him when he returned from a vacation in Monrovia. Now, three days later, his test results were not back. He looked healthy, and the medical team wanted to make sure he did not contract Ebola from the other patients. "They told me I should be very mindful of others. No touching," he said. His bed, like the others in the unit, was in an 8-by-10-foot space separated from others by wood-framed walls of tarp, and he shared a latrine with other patients.

One night, he said, someone died under the same roof. "It's too pathetic," Mr. Davies said through tears. "I think the world needs to come."

10 a.m.

Two men wearing yellow protective suits and thick rubber gloves left the clinic carrying a body bag on a stretcher. As they walked through tropical forest filled with birdsong, another man followed behind, spraying the dirt path until the brown leaves glistened with bleach solution. They were burying a 38-year-old man. In the final stages of his illness, he had left his bed, disoriented, and curled up against a 50-year-old woman who had died. A nurse found them lying together the next morning, a scene Dr. Hatch called "simultaneously touching and horrifying."

Now the man's body bag dripped as the burial team lowered it into a freshly dug hole, across from the grave of the woman, which was marked with a simple wooden sign. His was the seventh body on the grounds of the new treatment center. Ten more holes had already been dug, and four men with shovels stood by watching. No prayers were spoken, no tears were shed in what has become a new, numbed burial rite. As they returned, the team sprayed their path all the way back to the morgue.

10:50 a.m.

Doctors went over the first set of test results from a new mobile lab that the United States Navy had set up the previous day on the grounds of the local university. It seemed miraculous: test results would now be delivered in just a few hours instead of the four to five days required earlier, when samples had to be driven to Monrovia and were sometimes lost. Mr. Davies learned that he tested negative for Ebola and could leave.

Encircled by orange fencing, the center looks something like a detention facility. In truth it would be easy to escape; one confused patient had already wandered onto the driveway and been escorted back to the ward. No armed guards watch outside, and the entries to patient areas are not secured.

Another patient, Lorpu Kollie, 28, rejoiced after being told she was not on the list of those whose tests indicated infection. She called her parents to tell them she was going home. But a staff member had mistaken her for someone else; in fact her blood sample, drawn nearly a week earlier, had tested positive for Ebola.

Given the new diagnosis, Ms. Kollie was devastated and refused to move to the confirmed Ebola ward. Dr. Feuchte, the psychologist, went to meet with her. "I told everybody I am coming home," Ms. Kollie told her, crying. Dr. Feuchte said that she could stay put for now and they would test her again. A fellow patient sought to reassure Ms. Kollie, reminding her that she could be all right. Don't worry, she said, "only few more days."

Even with the new lab, something as basic as blood tests still posed outsize challenges. The intense disinfecting process sometimes erased or eroded patient identification numbers from tubes and paper, making them difficult or impossible to read. Each blood draw was risky for staff members, particularly because the syringes were difficult to manipulate with triple-gloved hands. "It would be inexcusable for us to have to take blood again and get a needle stick," Dr. Bucks had said during the morning shift change. Within hours he and others devised a better system for identifying tubes and discussed ordering a different blood-drawing system that would be safer to use at patient bedsides.

Noon

Lunch arrived; the heat was stifling. The blue buckets that served yam porridge at breakfast were now filled with rice. Workers set out food on plastic foam plates in the staff sitting room, waiting for doctors and nurses to wolf it down when they found a free moment.

1 p.m.

A staff member returned from the local market with bags of secondhand clothing, items to replace patients' clothes that were discarded because they were soiled or possibly contaminated. Among his haul: 20 lappas, local batik wraps worn as skirts; 20 bras; 30 pairs of socks; and piles of battered jackets and pants. "This is tenth-hand," joked Eric Diudonne, a civil engineer, who sent the clothing over to the laundry staff to wash.

The cleaners in the laundry room disinfected and washed the staff members' scrubs, goggles, rubber gloves, aprons, and boots in vats of diluted bleach and then ran them through a washing machine. The items were taken outside to dry, stiff aprons hanging from a line and boots propped upside down on poles. Some patients passed the time by washing their own clothing by hand, but all patient bedding and clothing was burned after they were discharged or died.

4:30 p.m.

Nurses gowned up to give afternoon medicines, run intravenous fluids, and draw blood from new patients and three who had to be retested. And the new results arrived. Ms. Kollie was now negative, meaning that she had recovered from Ebola or her first sample had been mixed up with someone else's. "Thank you, thank you!" Ms. Kollie rejoiced when she was told the news. She danced and reached out her arms as if to hug the staff, who were separated from her by plastic fencing. "I'm coming home," she told her parents on the phone.

She stepped into a shower for 10 minutes of scrubbing, with a nurse giving instructions. She dropped a handful of belongings, including money, a small change purse, and a cell phone, into a bucket of chlorine for disinfection. (She said she did not care that the phone might not survive.) Someone handed her a bag with a fresh set of clothing and $10 for transportation.

It had rained briefly. Looking out from the hilltop, Ms. Kollie saw a rainbow.

7:10 p.m.

The day team handed off the patients to the night crew. Several patients appeared to be doing better. And some whose tests turned out to be negative would be going home the next day.

But, a nurse reported, one man had vomited in the yard. Others had found a sheathed knife under his pillow, and he explained that he would rather die from a knife than from Ebola. Two days later, the disease killed him.

ATUL GAWANDE

No Risky Chances

FROM *Slate*

I LEARNED ABOUT a lot of things in medical school, but mortality wasn't one of them. Although I was given a dry, leathery corpse to dissect in anatomy class in my first term, our textbooks contained almost nothing about aging or frailty or dying. The purpose of medical schooling was to teach how to save lives, not how to tend to their demise.

I had never seen anyone die before I became a doctor, and when I did, it came as a shock. I'd seen multiple family members —my wife, my parents, and my children—go through serious, life-threatening illnesses, but medicine had always pulled them through. I knew theoretically that my patients could die, of course, but every actual instance seemed like a violation, as if the rules I thought we were playing by were broken.

Dying and death confront every new doctor and nurse. The first times, some cry. Some shut down. Some hardly notice. When I saw my first deaths, I was too guarded to weep. But I had recurring nightmares in which I'd find my patients' corpses in my house —even in my bed.

I felt as if I'd failed. But death, of course, is not a failure. Death is normal. Death may be the enemy, but it is also the natural order of things. I knew these truths abstractly, but I didn't know them concretely—that they could be truths not just for everyone but also for this person right in front of me, for this person I was responsible for.

You don't have to spend much time with the elderly or those with terminal illness to see how often medicine fails the people it is supposed to help. The waning days of our lives are given over to treatments that addle our brains and sap our bodies for a sliver's chance of benefit. These days are spent in institutions—nursing homes and intensive-care units—where regimented, anonymous routines cut us off from all the things that matter to us in life.

As recently as 1945, most deaths occurred in the home. By the 1980s, just 17 percent did. Lacking a coherent view of how people might live successfully all the way to the very end, we have allowed our fates to be controlled by medicine, technology, and strangers.

But not all of us have. That takes, however, at least two kinds of courage. The first is the courage to confront the reality of mortality—the courage to seek out the truth of what is to be feared and what is to be hoped when one is seriously ill. Such courage is difficult enough, but even more daunting is the second kind of courage—the courage to act on the truth we find.

A few years ago I got a late-night page: Jewel Douglass, a 72-year-old patient of mine receiving chemotherapy for metastatic ovarian cancer, was back in the hospital, unable to hold food down. For a week her symptoms had mounted: they started with bloating, became waves of crampy abdominal pain, then nausea and vomiting.

Her oncologist sent her to the hospital. A scan showed that despite treatment, her ovarian cancer had multiplied, grown, and partly obstructed her intestine. Her abdomen had also filled with fluid. The deposits of tumor had stuffed up her lymphatic system, which serves as a kind of storm drain for the lubricating fluids that the body's internal linings secrete. When the system is blocked, the fluid has nowhere to go. The belly fills up like a rubber ball until you feel as if you will burst.

But walking into Douglass's hospital room, I'd never have known she was so sick if I hadn't seen the scan. "Well, look who's here!" she said, as if I'd just arrived at a cocktail party. "How are you, doctor?"

"I think I'm supposed to ask you that," I said.

She smiled brightly and pointed around the room. "This is my husband, Arthur, whom you know, and my son, Brett." She got me grinning. Here it was, 11 at night, she couldn't hold down an ounce of water, and she still had her lipstick on, her silver hair was brushed straight, and she was insisting on making introductions.

Her oncologist and I had a menu of options. A range of alternative chemotherapy regimens could be tried to shrink the tumor burden, and I had a few surgical options too. I wouldn't be able to remove the intestinal blockage, but I might be able to bypass it, I told her. Or I could give her an ileostomy, disconnecting the bowel above the blockage and bringing it through the skin to empty into a bag. I would also put in a couple of drainage catheters—permanent spigots that could be opened to release the fluids from her blocked-up drainage ducts or intestines when necessary. Surgery risked serious complications—wound breakdown, leakage of bowel into her abdomen, infections—but it was the only way she might regain her ability to eat.

I also told her that we did not have to do either chemo or surgery. We could provide medications to control her pain and nausea and arrange for hospice care at home.

This is the moment when I would normally have reviewed the pros and cons. But we are only gradually learning in the medical profession that this is not what we need to do. The options overwhelmed her. They all sounded terrifying. So I stepped back and asked her a few questions I learned from hospice and palliative-care physicians, hoping to better help both of us know what to do: What were her biggest fears and concerns? What goals were most important to her? What trade-offs was she willing to make?

Not all can answer such questions, but she did. She said she wanted to be without pain, nausea, or vomiting. She wanted to eat. Most of all, she wanted to get back on her feet. Her biggest fear was that she wouldn't be able to return home and be with the people she loved.

I asked what sacrifices she was willing to endure now for the possibility of more time later. "Not a lot," she said. Uppermost in her mind was a wedding that weekend that she was desperate not to miss. "Arthur's brother is marrying my best friend," she said. She'd set them up on their first date. The wedding was just two days away. She was supposed to be a bridesmaid. She was willing to do anything to make it, she said.

Suddenly, with just a few simple questions, I had some guidance about her priorities. So we made a plan to see if we could meet them. With a long needle we tapped a liter of tea-colored fluid from her abdomen, which made her feel at least temporarily better. We gave her medication to control her nausea. We discharged her with instructions to drink nothing thicker than apple juice and to return to see me after the wedding.

She didn't make it. She came back to the hospital that same night. Just the car ride, with its swaying and bumps, made her vomit, and things only got worse at home.

We agreed that surgery was the best course now and scheduled it for the next day. I would focus on restoring her ability to eat and putting drainage tubes in. Afterward she could decide if she wanted more chemotherapy or to go on hospice.

She was as clear as I've seen anyone be about her goals, but she was still in doubt. The following morning she canceled the operation. "I'm afraid," she said. She'd tossed all night, imagining the pain, the tubes, the horrors of possible complications. "I don't want to take risky chances," she said.

Her difficulty wasn't lack of courage to act in the face of risks; it was sorting out how to think about them. Her greatest fear was of suffering, she said. Couldn't the operation make it worse rather than better?

It could, I said. Surgery offered her the possibility of being able to eat again and a very good likelihood of controlling her nausea, but it carried substantial risk of giving her only pain without improvement or adding new miseries. She had, I estimated, a 75 percent chance that surgery would make her future better, at least for a little while, and a 25 percent chance it'd make it worse.

The brain gives us two ways to evaluate experiences like suffering—how we apprehend such experiences in the moment and how we look at them afterward. People seem to have two different selves—an *experiencing* self who endures every moment equally and a *remembering* self who, as the Nobel Prize–winning researcher Daniel Kahneman has shown, gives almost all the weight of judgment afterward to just two points in time: the worst moment of an ordeal and the last moment of it. The remembering self and the experiencing self can come to radically different opinions about the same experience—so which one should we listen to?

This, at bottom, was Jewel Douglass's torment. Should she heed her remembering self—or, in this case, anticipating self—which was focused on the worst things she might endure? Or should she listen to her experiencing self, which would likely endure a lower average amount of suffering in the days to come if she underwent surgery rather than just going home—and might even get to eat again for a while?

In the end a person doesn't view his life as merely the average of its moments—which, after all, is mostly nothing much, plus some sleep. Life is meaningful because it is a story, and a story's arc is determined by the moments when something happens. Unlike your experiencing self, which is absorbed in the moment, your remembering self is attempting to recognize not only the peaks of joy and valleys of misery but also how the story works out as a whole. That is profoundly affected by how things ultimately turn out. Football fans will let a few flubbed minutes at the end of a game ruin three hours of bliss—because a football game is a story, and in stories, endings matter.

Jewel Douglass didn't know if she was willing to face the suffering that surgery might inflict and feared being left worse off. "I don't want to take risky chances," she said. She didn't want to take a high-stakes gamble on how her story would end. Suddenly I realized she was telling me everything I needed to know.

We should go to surgery, I told her, but with the directions she'd just spelled out—to do what I could to enable her to return home to her family while not taking "risky chances." I'd put in a small laparoscope. I'd look around. And I'd attempt to unblock her intestine only if I saw that I could do it fairly easily. If it looked risky, I'd just put in tubes to drain her backed-up pipes. I'd aim for what might sound like a contradiction in terms: a palliative operation—an operation whose overriding priority was to do only what was likely to make her feel immediately better.

She remained quiet, thinking.

Her daughter took her hand. "We should do this, Mom," she said.

"Okay," Douglass said. "But no risky chances."

When she was under anesthesia, I made a half-inch incision above her belly button. I slipped my gloved finger inside to feel for space to insert the fiberoptic scope. But a hard loop of tumor-caked bowel blocked entry. I wasn't even going to be able to put in a camera.

I had the resident take the knife and extend the incision upward until it was large enough to see in directly and get a hand inside. There were too many tumors to do anything to help her eat again, and now we were risking creating holes we'd never be able to repair. Leakage inside the abdomen would be a calamity. So we stopped.

No risky chances. We shifted focus and put in two long plastic drainage tubes. One we inserted directly into her stomach to empty the contents backed up there; the other we laid in the open abdominal cavity to empty the fluid outside her gut. Then we closed up, and we were done.

I told her family we hadn't been able to help her eat again, and when Douglass woke up, I told her too. Her daughter wept. Her husband thanked us for trying. Douglass tried to put a brave face on it. "I was never obsessed with food anyway," she said.

The tubes relieved her nausea and abdominal pain greatly—"90 percent," she said. The nurses taught her how to open the gastric tube into a bag when she felt sick and the abdominal tube when her belly felt too tight. We told her she could drink whatever she wanted and even eat soft food for the taste. Three days after surgery, she went home with hospice care to look after her.

Before she left, her oncologist and oncology nurse practitioner saw her. Douglass asked them how long they thought she had. "They both filled up with tears," she told me. "It was kind of my answer."

A few days later she and her family allowed me to stop by her home after work. She answered the door, wearing a robe because of the tubes, for which she apologized. We sat in her living room, and I asked how she was doing.

Okay, she said. "I think I have a measure that I'm slip, slip, slipping," but she had been seeing old friends and relatives all day, and she loved it. She was taking just Tylenol for pain. Narcotics made her drowsy and weak, and that interfered with seeing people.

She said she didn't like all the contraptions sticking out of her. But the first time she found that just opening a tube could take away her nausea, she said, "I looked at the tube and said, 'Thank you for being there.'"

Mostly we talked about good memories. She was at peace with God, she said. I left feeling that, at least this once, we had done it right. Douglass's story was not ending the way she ever envisioned, but it was nonetheless ending with her being able to make the choices that meant the most to her.

Two weeks later, her daughter Susan sent me a note. "Mom died on Friday morning. She drifted quietly to sleep and took her last breath. It was very peaceful. My dad was alone by her side with the rest of us in the living room. This was such a perfect ending and in keeping with the relationship they shared."

I am leery of suggesting that endings are controllable. No one ever really has control; physics and biology and accident ultimately have their way in our lives. But as Jewel Douglass taught me, we are not helpless either—and courage is the strength to recognize *both* of those realities. We have room to act and shape our stories—although as we get older, we do so within narrower and narrower confines.

That makes a few conclusions clear: that our most cruel failure in how we treat the sick and the aged is the failure to recognize that they have priorities beyond merely being safe and living longer; that the chance to shape one's story is essential to sustaining meaning in life; and that we have the opportunity to refashion our institutions, culture, and conversations to transform the possibilities for the last chapters of all of our lives.

LISA M. HAMILTON

Linux for Lettuce

FROM *Virginia Quarterly Review*

FROM A DISTANCE Jim Myers looks like an ordinary farmer. Most autumn mornings he stands thigh-deep in a field of wet broccoli, beheading each plant with a single sure swipe of his harvest knife. But under his waders are office clothes, and on his wrist is an oversized digital watch with a push-button calculator on its face. As his hand cuts, his eyes record data: stalk length and floret shape, the purple hue of perfect heads and the silver specks that foretell rot. At day's end his broccoli goes to the food bank or the compost bin —it doesn't really matter. He's there to harvest information.

Myers is a plant breeder and professor of genetics at Oregon State University. The broccoli in his field has a long and bitter story, which he told me last September at the university's research farm. We sat at a picnic table under a plum tree that had dropped ripe fruit everywhere; around our feet the little purple corpses hummed with wasps that had crawled inside to gorge on sweet flesh. Myers has dark hair and dark eyes that are often set behind tinted glasses. In public he rarely registers enough emotion to move the thick mustache framing his mouth. Still, as he talked about the broccoli his voice buckled, and behind those shadowy lenses his eyes looked hard and tense.

In 1966 a breeder named Jim Baggett—Myers's predecessor at Oregon State—set out to breed a broccoli with an "exserted" head, which meant that instead of nestling in the leaves, the crown would protrude on a long stalk, making harvest easier. The method he used was basic plant breeding: mate one broccoli with another, identify the best offspring, and save their seed for the next season. Repeated over decades by Baggett and then Myers, this process produced the broccoli in the field that day. The heads were so nicely exserted, sparrows used them as a perch.

Most classical plant breeders will tell you that their work is inherently collaborative—the more people involved, the better. Baggett had used versions of another broccoli called Waltham, released by the University of Massachusetts in the 1950s, as part of the foundation for his original exserted-head lines. Hoping to advance its evolution by letting others work on it, he and Myers shared their germplasm (an industry term for seed) with breeders throughout the United States. One recipient was the broccoli division of Royal Sluis, a Dutch company that had a research farm in Salinas, California. Through the channels of corporate consolidation, that germplasm ended up with the world's largest vegetable-seed company, Seminis, which in 2005 was bought by the world's largest seed company, Monsanto. In 2011, Seminis was granted U.S. Patent 8,030,549—"Broccoli adapted for ease of harvest"—whose basic identifying characteristic was an exserted head. More than a third of the original plant material behind the invention was germplasm that Baggett had shared in 1983.

As Seminis began previewing its Easy Harvest broccoli to the farm press in 2011, the company's lawyers began calling Myers, requesting more samples of broccoli seed. The patent they held covered only a few specific varieties that the company had bred, but now they were applying to patent the trait itself—essentially, any sizable broccoli with an exserted head. They needed the Oregon State plants for comparison to prove their invention was, in patent language, truly "novel."

Last August the examiner seemed dubious, writing, "Applicant is in possession of a narrow invention limited to the deposited lines; however, they are claiming any and every broccoli plant having the claimed characteristics." The application was given a "Final Rejection."

And yet, as Myers told me at the picnic table in September, "That's not necessarily final." Just before Thanksgiving, Seminis appealed, beginning a process that may last for years. As one intellectual-property manager who helps write patents for the University of Wisconsin told me, some examiners simply "cave and grant the broader claims as they get worn down by the attorneys' arguments." If Seminis receives the patent, their claim would likely encompass the plants growing in Myers's plots at Oregon State, meaning they could sue him for infringement.

Myers is not alone in this predicament. Irwin Goldman, a professor at the University of Wisconsin, had been developing a red carrot for 15 years when, in 2013, he learned that Seminis had an application pending for "carrots having increased lycopene content"—in other words, very red carrots. Likewise, Frank Morton, a small-scale, independent plant breeder in Oregon, had finally achieved a lettuce that is red all the way to its core, only to find that the Dutch seed company Rjik Zwaan had received a patent on that very trait. Their cases are just some of many.

When Myers talks about the issue, his frustration seems to turn him inward toward greater silence. But Morton is considerably less reserved. "It rubs me the wrong way that works of nature can be claimed as the works of individuals," he said, his voice growing louder and louder. "To me, it's like getting a patent on an eighteen-wheeler when all you did was add a chrome lug nut."

Myers contends that when applied to plants, patents are stifling. They discourage sharing, and sharing is the foundation of successful breeding. That's because his work is essentially just assisting natural evolution: he mates one plant with another, which in turn makes new combinations of genes from which better plants are selected. The more plants there are to mix, the more combinations are made, and the more opportunities there are to create better plants. Even some breeders who work for the companies that are doing the patenting still believe in—indeed, long for—the ability to exchange seed.

"It's this collective sharing of material that improves the whole crop over time," Myers told me. "If you're not exchanging germplasm, you're cutting your own throat."

If all of this seems like the concern of a specialized few, consider that plant breeders shape nearly every food we eat, whether a tomato from the backyard or the corn in the syrup in a Coke. Because of intellectual-property restrictions, their work increasingly takes place in genetic isolation and is less dynamic as a result. In the short term that can mean fewer types of tomatoes to plant in the garden or fewer choices for farmers and, by extension, consumers. In the long term it could hinder the very resilience of agriculture itself. Having access to a large genetic pool is critical for breeders who are adapting crops to the challenges of climate change. Every time intellectual-property protections fence off more germplasm, that gene pool shrinks.

What infuriates Myers, though, is that patents such as the one Seminis is seeking don't just impede sharing; they deter others from using their own germplasm. As the examiner noted, Seminis's patent application claims essentially all broccoli with an exserted head of a commercial size. If Myers's plants are too similar to those grown by Seminis, he won't be able to release his own variety for fear of patent infringement. Even if he did, no farmer or seed company would use it lest they be sued for the same violation.

"If they get the patent, they really hold all the cards," Myers said, wasps buzzing around his feet. "Then it comes down to at some point deciding whether to continue my program or to hang it up. Sell off the germplasm . . ." His voice trailed off. Then he gave a sad little laugh. The only buyer, of course, would be Seminis.

Fueled by both frustration and outrage, Myers, Morton, and Goldman helped establish a subtly radical group called the Open Source Seed Initiative (OSSI) in 2012. Operating under the radar, its mission was to reestablish free exchange by creating a reservoir of seed that couldn't be patented—"a national park of germplasm," Goldman called it. By 2013 the group had two dozen members, several of them distinguished plant breeders from public universities across the country.

OSSI's de facto leader is Jack Kloppenburg, a social scientist at the University of Wisconsin who has been involved with issues concerning plant genetic resources since the 1980s. He has published widely about the concept behind OSSI, and his words are now echoed (even copied verbatim) by public plant-breeding advocates in Germany, France, and India. As he explains it, for most of human history seeds have naturally been part of the commons—those natural resources that are inherently public, like air or sunshine. But with the advent of plant-related intellectual property and the ownership it enables, this particular part of the commons has become a resource to be mined for private gain. Thus the need for a *protected* commons—open-source seed. Inspired by open-source software, OSSI's idea is to use "the master's tools" of intellectual property, but in ways the master never intended: to create and enforce an ethic of sharing.

Kloppenburg's office plays to caricatures of lefty academics: every flat surface stacked with books and papers, a poster of Karl Marx on the wall. At OSSI meetings, amid a sea of plaid button-downs, he sticks out in his collarless, hemp-looking shirt. But he is fiery and, as one OSSI member says, "persistent as hell."

"The reason I'm doing this," he said, leaning forward in his creaking swivel chair, "is that I've spent the last twenty-five years doing the other thing, and what have we got?" That "other thing" has been exploring nearly every possible avenue to put control of seeds back in the hands of farmers and public-minded plant breeders: orchestrating international treaties, challenging interpretations of patent law, lobbying to amend the laws themselves —in other words, slow change. Indeed, over the course of three decades, it has felt to Kloppenburg like barely any change at all. Now nearing retirement, he wants action. He sees open source as a kind of end run. "The beauty of it," he said, "is that finally we get to create some space that is ours, not theirs."

As Kloppenburg talked about OSSI, he covered territory from the monopolistic tendencies of the American Seed Trade Association to Colombian peasant protests to the little-known story of German prisoners of war being used as forced labor in American corn-breeding fields. He pulled a hulking dictionary from the bookshelf and read aloud the precise definitions of *ownership* and *property*. He made it clear that while OSSI's practical goal was to create a reservoir of shared germplasm, its true mission was to redistribute power.

In this era of ownership, the consolidation of seed companies has meant the consolidation of control over germplasm, the industry's most essential tool. The plant breeders behind OSSI decry that trend for the constraints it puts on their individual breeding work, but they also see its damage in global terms. As founding member Bill Tracy, a sweet-corn breeder at the University of Wisconsin, articulated in his paper "What Is Plant Breeding?": "Even if we assume that the one or two companies controlling a crop were completely altruistic, it is extremely dangerous to have so few people making decisions that will determine the future of a crop . . . The future of our food supply requires genetic diversity, but also demands a diversity of decision makers."

*

People who sell seeds have always struggled with an inconvenient reality: their merchandise reproduces itself. In the past this has meant that farmers needed to purchase it only once, and competitors could make a copy by merely sticking it in the ground. In order for seeds to become a commodity and generate a profit, there had to be a reason for people to buy them year after year. Over the course of the 20th century, the industry devised certain solutions, including hybrid seeds and "trade-secret" protections for their breeding processes and materials. But perhaps the most effective solution is the application of intellectual-property rights, of which the utility patent is the gold standard.

More commonly associated with things like electronics and pharmaceuticals, the utility patent is a fortress of protection. It lasts for 20 years and allows even inadvertent violations to be penalized. Since the Patent Act of 1790, its intent has been to inspire innovation by giving exclusive rights to reproduce or use an invention, allowing its creator to reap a just reward. It was in exactly those terms that Monsanto's vegetable communications manager, Carly Scaduto, explained the Seminis exserted-head broccoli patent to me. "On average, it takes Monsanto vegetable breeders between eight and twelve years to develop and commercialize a new vegetable seed variety," she wrote. "Obtaining patents [is a way] for us to protect our time, ideas, and investment spent to develop those products."

It took seed companies nearly a century to secure that protection. As early as 1905 industry leaders advocated "patent-like" protection for plants, but they ran up against society's ethical resistance to patenting a product of nature. This view was famously aired by the United States Patent Office itself in 1889, in its denial of an application to patent a fiber found in pine needles. If it were allowed, the commissioner reasoned, "patents might be obtained upon the trees of the forest and the plants of the earth, which of course would be unreasonable and impossible." But many plant breeders insisted that their work was on a par with that of mechanical and chemical engineers. Their desire to achieve the same exclusive control over their inventions eventually led to the Plant Patent Act (PPA) of 1930. According to the committee report accompanying the Senate's version of the bill, the purpose was to "assist in placing agriculture on a basis of economic equality with industry . . . [and] remove the existing discrimination between plant developers and industrial inventors." Thomas Edison, already a household name for his own inventions, was enlisted to lobby for the bill and later lauded the PPA's passage to a reporter from the *New York Times*. "As a rule the plant breeder is a poor man, with no opportunity for material rewards," he said. "Now he has a grubstake."

What finally became law was in fact quite narrow. Instead of allowing utility patents for plants, the PPA created a new "plant patent," which applied only to plants reproduced asexually, like roses or apples, whose limbs are cloned. It excluded plants that reproduce sexually, through seed—which included wheat, corn, rice, and nearly every other staple food crop. The official reasoning was that sexually produced offspring weren't guaranteed to be identical replicas of the original plant—"true to type"—and so enforcement of a patent would be difficult. (It is notable, though, that an additional exclusion was made for tubers, which reproduce asexually but include potatoes—another indispensable food.) Writing in the *Journal of the Patent Office Society* in 1936, patent examiner Edwin M. Thomas explained the true reasoning: "The limitation, 'asexually reproduced,' was put in the law to prevent monopolies upon the cereal grains or any improvements thereof, while the limitation, 'other than tuber-propagated' was introduced to prevent patent monopolies on potatoes, etc." Congress had condoned the general concept of patenting plants, but it had drawn the line at patenting seeds of the sort that farmers plant and people eat.

By midcentury the official reasoning was moot. Advances in breeding had enabled seed producers to ensure that their plants would grow true to type, leading the industry to renew its efforts for protective legislation. Its first victory was the Plant Variety Protection Act, approved in a voice vote by a lame-duck session of Congress, on Christmas Eve, 1970. The act granted intellectual-property rights that were much like a patent, but it was tempered by concessions to those who continued to oppose the exclusive control an actual patent would have granted: farmers were allowed to save and replant seed from protected varieties, and researchers could use them in breeding their own plants. The real victory—the one the industry had been seeking for nearly a century—happened in 1980, when the U.S. Supreme Court ruled that life-forms could be patented if they were a new "composition of matter" produced by human ingenuity. That case concerned bacteria, but in 1985 the U.S. Patent Office extended the logic to plants. By the time this policy was affirmed by the Supreme Court in 2001, already 1,800 utility patents had been granted on plants, plant parts, and seeds.

The availability of this long-sought protection transformed the industry by solidifying the opportunity to treat seed as a proprietary technology. Already the promise of genetic engineering was attracting investment from international chemical companies and others whose experience lay more with developing industrial products than with breeding plants. Wielding this newfound, impenetrable intellectual-property protection, companies like Monsanto, Ciba-Geigy (now Syngenta), and Dow redesigned the business using a revolutionary metaphor: seeds were software. Genetics were improved almost surgically, with breeders altering DNA the way programmers rewrite code. The resulting corn, soybeans, and other commodities were modular components of a larger agricultural operating system, designed to work only with the company's herbicides. Even some labeling began to take a play from Microsoft: the seller's licensing agreement was printed on the back of seed bags in six-point font. Users didn't sign it; as with a box containing a copy of Microsoft Office, they agreed to it by simply opening the package. Among other things, those terms specifically prohibited use in plant breeding.

Market analyst Phillips McDougall calculated that in 1995, right around the time the software metaphor began to take hold, the global seed business was worth $14.5 billion. By 2013 it had grown more than 250 percent, to $39.5 billion. Transparency Market Research, which calculates a similar figure for 2013, forecasts the business will grow to $52 billion by 2018. In this context, the patent office's 1889 assertion that patenting the "plants of the earth" would be unreasonable and impossible sounds dated, if not naive. Seen through the lens of this new metaphor, patents make perfect sense. If seeds are software, then protecting them as intellectual property is a natural, even essential, requirement for their technological development. In a 2004 legislative study, the United Nations' Food and Agriculture Organization explained that this encouraged breeders "to invest the resources, labour and time needed to improve existing plant varieties by ensuring that breeders receive adequate remuneration when they market the propagating material of those improved varieties." In other words, innovation no longer grew out of sharing, it came from monopoly. "In the absence of a grant of exclusive rights to breeders," the report concluded, "the dangers of free riding by third parties would be considerable."

In 1997, as the laws of intellectual property had begun supplanting the ethic of sharing, a mild-mannered bean breeder named Tom Michaels also began thinking about seeds as software—but with radically different results. Michaels was struggling with the brave new world unfolding at his job in the University of Minnesota's horticultural sciences department. Until recently germplasm samples had simply been mailed between colleagues with no more than a friendly note, just as the exserted-head broccoli seed had been. But Michaels began to see this tradition of open exchange being curbed by legal documents that restricted research and demanded royalties. He tripped on the new vocabulary, which stipulated conditions about "unmodified derivatives" and "reach-through rights."

"If you're in plant breeding, you know you can't do it on your own," Michaels told me. "But I remember thinking, If this is the direction we're going, we all become islands. So what could we do to assure that we continued to work interrelatedly?"

During that time Michaels's computer-savvy son was messing around with alternative operating systems for his PC. Through him, Michaels learned about Linux and other software that was free to be used, altered, and shared by anyone. Linux came with a license that turned the concept of licensing on its head: instead of restricting people from copying the product, it restricted people from restricting it or any of its offshoots. It marked the code indelibly as part of the commons.

One fateful morning in Minneapolis, Michaels awoke with a Linux-inspired epiphany: *What if we did the same thing with our seeds?* Just like hackers, he and his colleagues would make their germplasm "free" by attaching a license that kept it in the public domain. No one could patent or otherwise restrict it or its offspring. Over time, Jack Kloppenburg and others heard about the idea, and together they honed it into the shrewdly elegant concept of open-source seed.

When Michaels first presented his idea to a group of fellow

bean breeders in 1999, it wasn't greeted as a grand prophecy. Jim Myers was in the audience then and recalls that while he and others found it interesting, they simply didn't feel a need for it. Intellectual property was on the rise, but utility patents were still rare in vegetable crops. There were, however, already more than 500 on maize and at least 250 on soybeans; today most germplasm of practical use for those plants is restricted as intellectual property, much of it by patents.

Because they comprise a smaller share of the world agricultural market, only recently have vegetables begun to attract the multinational investment and technological attention that commodities have had for decades. Also, because there are so many types of vegetables, and countless variations within each, they are much harder to blanket with intellectual property. Traded by gardeners around the world, vegetable seed still has a cultural identity—it is not yet simply software. Even within the industry, much of vegetables' breeding, and control of the germplasm, remains in the public sector.

Kloppenburg sees vegetables as the realm where open source can take root. "Corn and soybeans don't turn anybody on," he told me. "Nobody *eats* corn and soybeans. But they do eat what our breeders are doing." When he speaks with consumers about the open-source-seed concept, he asks them, "Do you want the same people who are breeding corn and soybeans to be making decisions about the stuff you buy at the farmers' market? Or do you want *Irwin's* beets and *Irwin's* carrots?"

That Irwin is Dr. Irwin Goldman, the University of Wisconsin vegetable breeder in patent limbo with his red carrots. If Kloppenburg is the brains behind OSSI, Goldman is the conscience, as warm and sincere as Kloppenburg is intense. When asked a question, he sits with his head of curly gray hair tilted to one side, neck thrust forward, in a posture of *really* listening. When he answers, he often begins with, "That's a great question."

Curiously, despite his role as a founding member and unofficial vice president of OSSI, Goldman holds three utility patents on vegetables—two on beets, one on carrots. He explains that the patented vegetables are used to create industrial dyes and have little crossover with food plants. Plus, it was the university that sought the patents in his name. Still, Goldman offers the disclosure like a personal confession. His explanation for going along with it is that he was young and foolish, a new professor seeking tenure. At the time his only reference point was his grandfather Isadore, a poor Russian immigrant who had designed and managed to patent a unique barber coat that didn't collect hair in its pockets. His family had always been deeply proud of Isadore. When Goldman found himself listed as the inventor of those beets and carrots, he flushed with the honor of this parallel achievement.

"But over time," he told me, "the experience of doing it made me realize what the implications of patents like those are. I asked myself, What would make me feel like I had made a contribution to the future—to a sustainable future?" After a hiatus during which he served as the college's dean, he returned to breeding and devoted the rest of his career to developing germplasm that is "free and clear."

Goldman agrees with Kloppenburg that vegetables are the most likely arena for OSSI to come to life. In his more hopeful moments, he envisions a food label alongside "organic" and "fair trade" that tells consumers their food is "open source." But, he warns, if they are going to claim any significant amount of genetic territory, OSSI needs to act fast. Patents already cover everything from "low pungency" onions to "brilliant white" cauliflower, and a gold rush is taking place, with seed companies scrambling to claim what territory remains. Since 2000, lettuce alone has garnered more than 100 patents; an additional 164 are pending. When Goldman went online to show me Seminis's red-carrot application, his search brought up another, newer application for a different red carrot that he hadn't even known about. During the writing of this article, seven more applications for patents on carrots have been filed.

"Open source still has a chance with vegetables, but our window is only as long as the bottleneck at the patent office," Goldman said. "It could be a matter of less than a decade before what has happened with corn happens with crops like carrots and onions."

On a sunny August day, at a research station in Mount Vernon, Washington, the men and women of OSSI were arranged around a flotilla of conference tables. The group was almost comically homogenous in appearance: two dozen men with gray hair, glasses, and collared shirts; a dozen women, young and athletic, mostly graduate research assistants. Kloppenburg sat at the head of the tables in a linen shirt and a turquoise necklace. Goldman was at his side.

The group had convened in order to finally transition open-source seed from a clever idea to a legally defensible system. They were all clear on the basic principle—that, as Kloppenburg has written, "the tools of the master are repurposed in a way that . . . actively subverts the master's hegemony." But an hour into determining exactly how to do that, eyelids were drooping. The coffee machine began gurgling out refills. "OSSI has indeed found," Kloppenburg would later write, "that the tools of the master are technically very cumbersome."

A sweet-corn breeder named Adrienne Shelton made the case that the "political jujitsu" of open-source software wouldn't work for seeds. When computer code is written, she explained, the author automatically gets copyright. That ownership allows the author to then take out a copy*left* that says the material can be used freely. But plant breeding isn't governed by copyright law, and by breeding a plant one does not automatically own it. One would need to patent the plant first in order to then claim the "patent left" of declaring it open source. "Most of the people that would be supportive of what we are trying to do as open source," Shelton said, "probably would be very, very skeptical if we said, Well, first we have to patent it."

An alternative would be to employ another of the master's tools: contract law. No patent would be necessary. Instead, before receiving germplasm, a person would sign a license agreeing to the open-source rules. On the table in front of Kloppenburg lay a draft of such a license, but no one could suffer the legalese long enough to survive even the first page in that cold pile of paper.

Goldman tilted his head and looked at the license with concern. "I can't imagine handing over a vial of seed and, oh, let me go to the copy machine and give you this seven-page, single-spaced document," he said. "It seems incompatible with what we're trying to do: the open seed, and then a license that if you want to understand, you need to ask your attorney."

Discussion turned to the quick and dirty "bag tag" licenses modeled on the stickers that sealed boxes of software; by opening the box or bag, the user agrees to the terms. Could a similar mechanism be used to mark seed as open source? Would it be legally binding? No one was sure.

Kloppenburg directed the group's attention to a series of slides on the screen behind him. They were advertisements for private security firms and other organizations that enforce plant-related intellectual-property rights in the United States, Europe, and South America. Many of the largest seed companies are partners, as are numerous land-grant universities, including the one where this meeting was being held. The Farmer's Yield Initiative, or FYI, offered a toll-free hotline where callers could submit anonymous tips about people using patented seed illegally.

Heads shook in disbelief and disgust, but the point had been made: intellectual-property protections work because of deterrence; the ill-fitting metaphor of seeds as software was held in place by fear. None of the OSSI members I asked was able to name a plant breeder who had been sued for patent infringement or broken contracts, and yet nearly every one of them was willing to abandon material he or she had been working on for years rather than test how forgiving the intellectual-property holders might be. Later Bill Tracy, the sweet-corn breeder, put it bluntly: "If you fear the company, you're not going to cross it and the patent works. If you don't fear the company, it doesn't work. It comes down to who has the most lawyers."

Looking around the room, it was clear this was not the group with the most lawyers. They had had one, who drafted their open-source license pro bono. But the week before, she had stopped returning their calls.

After the meeting I spoke with Andrew Kimbrell, a public-interest lawyer and the executive director of the Center for Food Safety. He has led numerous legal challenges to plant patenting, and he certainly sympathizes with OSSI's intentions. "In the midst of climate disruption," he said, "having a diverse seed supply created through a robust public breeding program is a food security and national security issue. For that alone we should get rid of this patent issue and invest in public plant breeding."

He advocated slower kinds of change: legislation to return to the days when farmers and plant breeders were free to use any seed as they wished; more legal challenges to puncture the precedent that leads courts to rule consistently in favor of intellectual-property protections. He even encouraged the basic, boring act of publishing research on plant breeding, since the most effective way to prevent something from being patented is to have documented that the thing already exists.

But the jujitsu that OSSI was trying to pull off he found "problematic" at best. "Just because you declare something open source doesn't mean it's off limits," he said. "It could simply mean that you passed up your chance to get to the patent office."

In the following months Kloppenburg, Goldman, and a few others began meeting weekly to try to salvage the idea and launch it, somehow, before another growing season slipped by. They spoke to half a dozen lawyers, who confirmed that the licenses wouldn't work. They were advised to patent their seed. "I never wanted to hire lawyers," Goldman told me, exasperated. "I don't want to be in the business of tracking licenses. I just want to free the seed."

The reason OSSI stumbled in trying to emulate computer programmers and open-source its seed wasn't just naïveté with legal matters. In a way, the larger problem was the metaphor itself. Seeds are not software; they are living entities that grow and reproduce. Indeed, that's the reason why the industry sought intellectual-property rights in the first place. But those protections can't truly contain biology—seeds slip right through barriers made of words. If you want to reproduce a patented soybean, just lift one from a farmer's field at harvest time and plant it in a pot. Without deterrence, a plant-utility patent is just an expensive piece of paper.

Even with a fleet of lawyers, chances are OSSI could never outsmart the intellectual-property system: normally patents and licenses need to last for only one generation of plants; they say the seed can't be planted back, and that's that. But open source was supposed to allow the material to proliferate, which means OSSI would need to make sure that its license accompanied every new generation of plant—an exponentially expanding demand. Enforcing that viral replication would be nearly impossible. Without it, the seed would go right back to the unprotected commons, where anyone could claim it and patent it. The fluid nature of seeds, their natural impulse to regenerate, is both the impetus for the open-source concept and its legal undoing.

In January the group drew up a new license. This time they dispensed with the legalese altogether and instead wrote from their hearts. At just three sentences long, it wasn't much of a legal document; it would never stand up in court. Instead they would print it on the outside of each packet, just as Seminis does theirs, but with the opposite effect. "This Open Source Seed pledge is intended to ensure your freedom to use the seed contained herein in any way you choose, and to make sure those freedoms are enjoyed by all subsequent users," it read. "By opening this packet, you pledge that you will not restrict others' use of these seeds and their derivatives by patents, licenses, or any other means."

Goldman toyed with the idea of also printing the pledge on slips of paper to be included inside the packet, like fortunes in a fortune cookie, to encourage people to pass it along. "I'm coming to see it more as a performance-art piece," he told me brightly.

Despite his optimism, the group was admittedly disappointed. The goal had been to replace their defensive stance around intellectual property with a legal mandate. "Instead of just saying, Oh, please don't patent these things, it's not right," Kloppenburg said, "we wanted a commons protected by law." Now they were back to relying on the thin armor of ethics and morality for protection. They were back to slow change.

But with a blue-sky tone, Kloppenburg said that the true objective had never been to create a license per se. Instead it had been to create a positive alternative to the intellectual- property regime. He was confident they could still do as much. The open-source idea had generated enthusiasm from all corners of the agricultural world. (Even the inventor of the Seminis red carrot, an old-school plant breeder caught in the tangle of the modern industry, had expressed his support.) This signaled to Kloppenburg that perhaps finally there was enough momentum to build an American seed movement big enough to have an impact.

OSSI was one of many nuclei organizing around the larger topic of seeds. Over the previous two years the national seed-swapping nonprofit Seed Savers Exchange had grown its membership by 33 percent—to 13,000 gardeners. Later, in 2014, Vermont would pass the first law in the country requiring the labeling of all foods containing genetically modified ingredients, a goal anti-biotech activists had sought for years. But there was still no concerted, sustained effort around that most fundamental issue of control and ownership, as there was in other countries. Kloppenburg pointed to Canada, where the National Farmers Union had been waging war against increased intellectual-property protections, and to countries throughout the developing world, where seed issues were an integral part of the international struggle for peasant rights. Across the Atlantic there had been an uproar for more than a year over European Patent 1,597,965: "Broccoli type adapted for ease of harvest"—granted to Seminis in May 2013. As the title suggests, the claim is essentially identical to the company's American patent on exserted-head broccoli. But while Jim Myers was about the only person upset about the U.S. version, a coalition of 25 organizations from Europe and India filed a formal opposition to the European patent within months of its approval. Along with the requisite paperwork requesting that it be revoked, they delivered 45,000 signatures from supporters.

"Patents on naturally occurring biodiversity in plant breeding are an abuse of patent law," the opposition statement read, "because instead of protecting inventions they become an instrument for the misappropriation of natural resources."*

Their argument centers around a single line in the European Patent Convention, Article 53b, which states that patents shall not be granted on "plant or animal varieties or essentially biological processes for the production of plants or animals." Recent objections to similar claims (one on a different broccoli, another on a tomato) led the European Patent Office's board of appeals to clarify that a new variety created by simply crossing plants and selecting their offspring—exactly the work of Myers and the Seminis broccoli breeders alike—was considered essentially biological and so not patentable.

The latest American ruling on the topic, in June 2013, established just the opposite. In the highly publicized case *Association for Molecular Pathology v. Myriad Genetics,* the U.S. Supreme Court ruled that DNA itself was "a product of nature and not patent eligible." But in delivering the opinion, Justice Clarence Thomas made the distinction that new plant breeds developed by conventional plant breeding *were* patent eligible. He cited the American Inventor's Protection Act of 1999 as well as court precedent—namely, the opinion in the landmark 2001 case, which he also authored.

* Opposition was filed also by Syngenta, the Swiss biotech company and direct competitor to Monsanto. Their objection followed the same general logic as the coalition's: that the broccoli under protection was created by an "essentially biological process." It was ironic, then, that they had just applied for their own patents in the United States and the European Union, covering a broccoli plant distinguished in part by a "protruding" head that makes harvest easier.

So while OSSI believes in the same basic principles as the European coalition—whose statement about the ethical implications of patents could have been written by Kloppenburg himself—the Americans' fight is arguably much tougher. Its challenge is to amend patent law, which involves lobbying Congress against the powerful forces that are deeply invested in maintaining, if not strengthening, intellectual-property protections. Slow change, indeed.

Kloppenburg hopes that OSSI, with its new approach, can at least help speed things up. Listening to him and Goldman describe their new vision, it's almost as if they have replaced seeds as software with a new metaphor—one inspired by plant breeding itself. Instead of building a protective barrier, OSSI would reach out into the world as widely as possible. Each time open- source seed was shared, the message on the packet would germinate in new minds: it would prod the uninformed to question why seeds would not be freely exchanged—why this pledge was even necessary. It would inspire those who already knew the issues of intellectual property to care more and spread the word. As the seed multiplied, so would the message. With three simple sentences, OSSI would propagate participants in the new movement like seedlings. They would breed resistance.

On April 17 the Open Source Seed Initiative announced itself to the public in a ceremony at the University of Wisconsin. The original plan was to rally on the steps of grand Bascom Hall, next to a bronze statue of Abraham Lincoln—"an appropriate witness to our emancipation of seed," Kloppenburg said. Instead the rally took place outside the less charismatic Microbial Sciences Building, beside a tree still bare in the young springtime. Unfazed, volunteers planted the dry lawn with dozens of short white flags reading free the seed!!!, which shivered in a brisk breeze. Clad in winter jackets, about 60 people gathered to hear Kloppenburg, Goldman, and others talk about food sovereignty and the importance of genetic diversity. Then organizers handed out packets printed with the OSSI pledge. Each contained seed from one of 36 open-source varieties, ranging from barley to zucchini. They included two carrots bred by Goldman, one of which he named Sovereign, in honor of the occasion.

They also included a broccoli from Oregon whose history began in 1997, the same year as Tom Michaels's epiphany about the future of plant breeding. That year Jim Myers began breeding a plant he now calls "The O.P.," which stands for "open-pollinated." Until then his broccoli were either hybrids or inbreds, created by a process of narrowing the genetics until one select mother is bred with one select father to create a single, most desirable combination of genes. The O.P., by contrast, is the result of a horticultural orgy. Myers began with 23 different broccoli hybrids and inbreds, including some of the lines behind the exserted-head trait. He let insects cross-pollinate them en masse, and the resulting plants were crossed at random again—and again, and again, four generations in a row. He then sent germplasm to farmers around the country, had them grow it in their fields and send back the seed they collected. Over the winter Myers bred it in another greenhouse orgy, then sent it back to farmers. For six years he repeated this process.

The broccoli evolved in two ways simultaneously. The back-and-forth of the breeding scrambled the plants' genetics, making the germplasm wildly diverse. It also let the environment whittle away at individual genes. For instance, plants without pest resistance produced less seed or simply died, reducing their presence in the gene pool. When it was hot, plants that could tolerate heat produced more seed, increasing their presence. Survival of the fittest.

In the seventh year Myers sent most of the seed back to the farmers—just gave it to them, without licenses, royalties, or restrictions. The idea was that each farmer would adapt that dynamic gene pool to his or her farm's particular climate and conditions, selecting the best plants every year to refine the population. In other words, they could breed it themselves. In time each would end up with his or her own perfect broccoli.

The beauty of the O.P. is that rather than challenge the intellectual-property system, it inherently rejects the concept of ownership. It contains many of the desirable genetics of Myers's commercial broccoli lines, but in a package that is designed to be shared, not owned. Because it is open-pollinated, not a hybrid, its seeds can be saved by any farmer. And because it is genetically diverse, it would be difficult to pin down with a patent. Even if someone did claim to own it, because each new seedling is a little different, that claim would be all but impossible to enforce. In this case, the plant's natural instinct to mate, multiply, change—to evolve—isn't an impediment at all. Rather, it is a central reason why people would want to grow it in the first place.

One of the farmers who received seed from Myers was Jonathan Spero, who grows and breeds vegetables on his farm in southwest Oregon. After a decade of working with the O.P., he released his own variety, a sweet, purplish broccoli that sends out numerous side shoots after the main head is harvested. Spero named it Solstice because it produces earlier than most—if planted by mid-April, it will yield florets by the first day of summer. Some people also refer to it as Oregon Long Neck, because it has an exserted head. On April 17, in front of the Microbial Sciences Building at the University of Wisconsin, it gained another title of sorts: the world's first open-source broccoli.

That day, as the last act in the ceremonial birth of OSSI, the audience turned over their crisp little packets of seed and recited the pledge on the back. What they held in their hands was no silver bullet. It wouldn't keep Seminis from getting its broccoli patent application approved, much less rewrite the laws of intellectual property in favor of free exchange and genetic diversity. But still, as Kloppenburg and Goldman read those precious words in unison with the small crowd gathered in the cold before them, there was a new power to their voices. The seed wasn't even in the ground yet, but already open source was taking root.

ROWAN JACOBSEN

Down by the River

FROM *Orion*

ALTHOUGH YOU WOULDN'T have known it as recently as 10 years ago, the Sonoran Desert city of Yuma, Arizona, is a river town. Located near the junction of California, Arizona, and Mexico, this kiln-dry city of 90,000 people and 3 billion heads of lettuce has always owed its existence to the Colorado River. It was here in 1849 that thousands of Gold Rushers arrived at Yuma Crossing, where two granite ledges funneled the powerful Colorado River through a deep narrows that made for the easiest ferry crossing in the Southwest. The native Quechan enjoyed a brief monopoly ferrying settlers across the Colorado until the U.S. Army established Fort Yuma in 1850, ostensibly to protect the river crossing. Regular skirmishes between the Quechan and the newcomers followed. Soon enough, more than a few Quechan were dead and the river crossing was firmly in the hands of the United States. The Quechan were relegated to the Fort Yuma Indian Reservation on the west bank, while the city of Yuma grew up on the east bank, welcoming riverboats that steamed up from the Gulf of California laden with settlers and supplies.

To picture the Colorado River back then, one needs plenty of imagination, because the current river in no way resembles the historical one. Before the coming of the dams, the Colorado was an unruly god of creation and destruction. Fed by Rocky Mountain snowmelt, it would swell to 40 times its size each spring, inundating the floodplain and forming one of the world's largest deltas, which ran from Yuma to the Gulf of California, 100 miles south. The floods were the lifeblood of this bone-dry corner of the country. Verdant gallery forests of willow and cottonwood etched the riverbanks, and thick mesquite groves filled the floodplains. The trees evolved to survive on one big dose of water per year by drilling their roots deep to reach the groundwater. Cattails and bulrushes thrived in the meanders and back channels, making the region the unlikeliest of ecological jewels. Nearly 400 species of birds used the desert wetlands. Beaver, deer, and pumas slipped through the cool green pools.

As did the Quechan. They fished and hunted year-round, made their houses from willow withes, cooked mesquite-flour tortillas over mesquite fires, and waited until the floods receded each spring to plant their crops in the still-damp earth.

That way of life didn't work for the sedentary settlers, who in 1936 were only too happy to see Hoover Dam shackle the beast that had eaten their homes and farms one time too many. More dams and canals followed, and the Southwest's great boom was on. By the time the Lake Powell Reservoir, about 700 miles upriver on the Arizona-Colorado border, filled in 1980, 90 percent of the river's water was being diverted to farms, as well as cities as far away as Denver and Los Angeles, before it reached Yuma. The mighty river had become an obedient stream that never left its banks, let alone overflowed into what had been desert wetlands. Deprived of their annual dose of river water, the willow, cottonwood, and mesquite trees died. The marshes withered. And the birds disappeared. The Pacific Flyway, which stretches from British Columbia to South America, became a shadow of its former self. The Quechan found themselves on a barren reservation.

One of the few plants that flourished in the arid and saline soils of the unwatered Lower Colorado floodplain was *Tamarix ramosissima,* the tamarisk or saltcedar, which looks like a ratty pine tree and forms tangled stands a few yards high. The tamarisk—described in both the Bible and the Koran and a native of the deserts of North Africa and the eastern Mediterranean—was planted by the millions during the Dust Bowl to fight soil erosion. With little competition from native plants, it soon swarmed across the West, and now infests more than 3 million acres, including almost the entire Colorado River corridor.

Few areas were hit harder than Yuma, and the calamity went beyond the tremendous loss of biodiversity. In 1999 community developer Charlie Flynn took the helm of the Yuma Crossing National Heritage Area, which is part of the National Park Service's program to foster community-driven stewardship of important natural or cultural landscapes. His task was to bring the riverfront back to life, but he found the area so overgrown with invasive tamarisk thickets that no one could get near the water, and in the few places where people could, they didn't dare because of drug smugglers who used the abandoned waterway as a thoroughfare. "Once all the nonnative vegetation grew up, it was the perfect breeding ground for drug traffic, meth labs, hobo camps, trash dumps," Flynn explained to me. "You name it, it was down there. It was a no-man's land. People just didn't go to the river. They were afraid to. Even the police hated going down there. You couldn't see two feet ahead of you."

Along with the riverfront, the historic downtown had withered. And with the two sides of the river in a cold war of sorts, the only direct connection between downtown Yuma and the Quechan reservation—the Ocean-to-Ocean Bridge—had been allowed to fall into ruin. The 1915 masterpiece was declared structurally deficient in 1988, and there had been no attempts to repair it. Yuma soon forgot about its river and its history and became a sprawling land of industrial lettuce farms fueled by migrant labor.

A practitioner of Eastern medicine might have diagnosed a chi blockage. The energy of the river was what had always held the system together. The tamarisk and the drug trade were the diseases that had taken hold in the resulting imbalance; the severing of community and the loss of identity were some of the symptoms. But with every drop of the Colorado River already claimed, and the Southwest projected to have 20 million additional people and even hotter temperatures in the coming decades, the river was unlikely to ever again have a surplus of water. Faced with these realities, the city shrugged and turned its back on its benefactor.

Charlie Flynn saw the upside of a restored waterfront, but he thought the odds of success were slim. "I wondered if it was really possible," he admitted to me. "The physical and stakeholder challenges were so great." The land-ownership issues were even more tangled than the tamarisk: 16 different entities owned land along the river, including the Quechan, the city of Yuma, local farmers, and the states of California and Arizona. And they would all have to work with the Bureau of Reclamation, which manages infrastructure on the Colorado River. "The Quechan didn't trust us. The farmers didn't trust us. You couldn't ask for a mix more hostile to the federal government."

Although the city and the tribe were not on speaking terms, Flynn knew nothing could proceed without the cooperation of the Quechan, so he took it upon himself to start attending the monthly tribal council meetings. "I thought, okay, is this going to be a transaction or a relationship? What most people don't understand is that if you start off trying to do a transaction with a Native American tribe, you're dead. You'd better establish a relationship, establish trust, and then go forward." At the first meeting he attended, he received a litany of injustices visited upon the tribe. "It was a monologue, going all the way back." But the monologue slowly turned into dialogue, and eventually Flynn and the tribe found two things they both wanted to fix: the riverfront and the Ocean-to-Ocean Bridge. "We agreed that the bridge was more tangible and doable. It was a rallying cry. And it was also a physical and symbolic connection between the Quechan and the city." After long negotiations, they agreed to split the costs evenly.

On February 28, 2002, the Ocean-to-Ocean Bridge reopened in a spirited celebration. About 800 people turned out to watch the mayor and the Quechan tribal president drive across the bridge in antique cars, and the Quechan hosted the ceremony on their side of the river. "The bridge was the turning point," Quechan economic development director Brian Golding told me. "Because of that collaboration, the two entities came together."

Two weeks after the bridge ceremony, another chasm was crossed when the Quechan tribal president called Charlie Flynn and proposed that they get to work on the riverfront. Flynn learned of a young landscape architect who had been living in Parker, Arizona, and restoring wetlands on the Colorado River Indian Reservation. So he drove about 100 miles north of Yuma to visit.

The Colorado River Indian Reservation is a dusty moonscape of cactus and red rock. No trees, no grass. In summer it cooks along at 110 degrees, and you can almost hear the crackle of creosote bush leaves frying in their oil. When Flynn arrived, a baby-faced man with a long ponytail introduced himself as Fred Phillips. Flynn found him to be practical and passionate about his work. "Although he had deep respect for Native Americans, he did not romanticize them," Flynn said. "He was willing to engage with tribal leaders, which can be both difficult and meaningful."

Phillips took Flynn on a tour of his restoration work. They drove down a long dirt road through the reservation's arid beige lands until they reached a riverbank. Flynn could hardly believe his eyes: hundreds of acres of green river marsh stretched before him. Two and a half miles of channels, fringed with rushes and waterfowl, flowed to the main stem of the Colorado. There were four miles of trails and a mesquite-filled park with a swimming area and boat ramp. Phillips assured Flynn that yes, the area had been just as tamarisk-clogged as Yuma when he started, and yes, what had happened here was replicable up and down the Lower Colorado.

Six years earlier Fred Phillips had been a junior at Purdue University, getting his degree in landscape architecture. He had a fascination with the native cultures of the Midwest river valleys and a growing disillusionment with college. "What I was being taught was how to create these artificial landscapes, which was not what I thought landscape architecture should be. I thought it should be about working with the land instead of imposing our will on it." When he learned through the grapevine that the Colorado River Indian Tribes wanted to restore their riverfront, he drove cross-country to the reservation to offer his services. An elder took him to a backwater that was filled in with tamarisk. The elder said, "I used to come down here with my dad, a sack of flour, and some bacon, and we'd hunt. There were big cottonwoods, and the river flowed through them. Now it's all gone. I'd really like to bring some of that back."

Phillips had no idea where to start. "I knew how to put a plan together, and I knew how to draw and design landscapes, but that was for residential properties and shopping malls. Master planning a thousand acres of restoration? No idea. So I just started gathering information. I talked to everyone from tribal elders to the Bureau of Reclamation. What did this area used to be? What are the problems? What are your ideas? I got some aerial photos. I started hiking and canoeing the whole thing. I was sure I was going to die of a rattlesnake bite."

Instead Phillips learned one of the key lessons for the Colorado River Basin. One often assumes that restoring natural landscapes is a long, gradual process, but the Colorado River ecosystem is different. Its native plants evolved amid annual disturbances, and they know how to move fast. Give them some free ground and a little bit of water and they go for it. The natural communities of the Lower Colorado were degraded, but perhaps the gap between a desiccated system and a flourishing one could be bridged more easily than it seemed. Just find the water, reset the conditions, and let the system heal itself.

While camping in his Airstream and working 70-hour weeks, Phillips devised and implemented a 250-page master plan. He reconnected an old meander to the river by dredging it, ripped out the tamarisk, and planted native species and kept them on drip irrigation until they could extend their roots to the water table. Amazingly, when wet, the cottonwood roots grew an inch a day.

Phillips also mastered the delicate art of permit applications. In addition to the tribes and the state, he had to convince the Bureau of Reclamation, the Army Corps of Engineers, the U.S. Fish and Wildlife Service, and the Bureau of Indian Affairs that this was a good idea. He learned to move between worlds.

By the time Charlie Flynn tracked down Phillips, he had restored 225 acres of wetlands along the river. His cottonwoods were 35 feet high. He had built parks and even started a nursery when he couldn't find a source of seedlings. The timing was perfect for Flynn and Yuma: Phillips had perfected his techniques and was itching to see what he could do on the rest of the river.

Fast-forward 10 years to a halcyon winter evening in 2014. Fred Phillips and I carry two paddleboards down to Gateway Park, on the riverfront below the Ocean-to-Ocean Bridge, and launch ourselves up the Colorado River. From our watery vantage point we can see riverbank trails full of bikers and birders. It's impossible to believe that this exact spot has been ground zero for various crimes against nature and humanity; the place is almost mundane in its pleasantness.

A few hundred yards upriver we portage over the riverbank and slide our boards into a back channel Fred has excavated. The sounds of people and roads fall away, and we find ourselves in a sea of cattails. Ducks and white-faced ibis burst from the marsh as we paddle by. Egrets high-step in the shallows and turn orange-pink in the low sunlight.

Four hundred acres of the Yuma wetlands have been restored with native species so far; another thousand are planned for restoration. Bird diversity has doubled, and even the Yuma clapper rail, one of the most endangered birds in the Southwest, has returned to the Yuma marsh. Water was the key, but the city of Yuma had little to spare. Fortunately, wetlands in the Sonoran Desert are not choosy or proud. Phillips rerouted runoff from neighboring lettuce fields and found another 325,000 gallons a day that the City of Yuma used to rinse the sand filters in its water treatment plant. The wetlands drank it up.

In coming years a garland of green is expected to sprout up and down the Lower Colorado, a beacon to birds and snowbirds alike —all part of a growing realization that far from being antithetical, ecological and cultural restoration reinforce one another. In addition to the Yuma wetlands, Phillips is working on a 1,000-acre restoration project a few miles upriver and a smaller one downriver along the Mexican border. Phillips's counterparts in Mexico have several more projects in the works and are using the techniques he pioneered.

Returning to the main stem of the river, Phillips and I visit Sunrise Point Park on the Quechan side, where a medicine wheel and ramada frame the summer solstice sunrise. Traditional Quechan herbs buzz with wild bees. Phillips picks a bundle of white sage, inhales deeply, and holds it to my nose. "Best kind for smudging," he says. The usual camphor aromas are leavened with a lemony brightness like a brain tonic. In a nearby clearing in the willows sits the Elder Village, a cluster of wattle-and-daub huts around a fire pit.

"Our people are finding their way back to the river," Brian Golding, of the Quechan tribe, later told me. "Culturally, spiritually, and physically, we're more connected. That continues to manifest itself. Our Elder Village was built by the community. We have our weddings down there, and our Youth Cultural Festival was just held down there."

Golding also sees potential economic development in the healthy wetlands. "We've got this veritable warehouse of material that can be used by elders and youth and middle-agers like myself for our cultural production needs. It's not just pretty to look at and attractive to birds; it's what we use for traditional structures, musical instruments, basketry, and artwork. We recently allowed a neighboring tribe to harvest some willow poles so they could make a sweat lodge. It feels good that we're able to provide that resource to folks who otherwise wouldn't be able to do it, or would have to use some other material that isn't authentic. Before, you couldn't get to it, or it was in such short supply that you didn't dare take any."

Yuma too has found its way back to the river. The adobe walls of the Yuma Territorial Prison, where many an outlaw did his time, rise on a bluff across the river. Farther down the waterfront is the Quartermaster Depot, where the riverboats used to unload. Both had been neglected state parks for years. In 2010 the cash-strapped state announced that it would be closing the parks, but the Yuma Crossing National Heritage Area raised $70,000 from the local community in 60 days, and the city was able to lease the parks from the state. The Heritage Area has run them ever since, and the fundraiser is now an annual event. The Quartermaster Depot has become the city commons Yuma lacked. It hosts a new farmers' market, running races, and the annual Lettuce Days festival, and its displays celebrate the city's historical and contemporary ties to the Colorado River. The prison, meanwhile, offers breathtaking views over the East Wetlands, with the silvery river snaking through and curling away into the desert.

The leafy oasis was the key draw for a new Hilton Garden Inn and conference center on the waterfront. The hotel promotes its location "on the banks of the majestic Colorado River in downtown Yuma, Arizona." And the Yuma River Tubing shop recently opened its doors beside the wetlands as well. All of this would have been unfathomable not long ago. In 2014 the National Park Service published an evaluation of the impact of National Heritage Areas and pulled no punches when describing the Yuma of old. The Park Service called the waterfront a "blight" and pointed out that the city had not been welcoming to the idea of bike or walking paths. "For decades," it said, "this region struggled to gain a sense of identity." Now, having rediscovered its river-town soul, Yuma's identity seems solid. The Park Service estimates the annual economic impact of the restored waterfront at $22.7 million, most of which comes through tourism. What it can't put a number on is how different it feels to be a Yuman, or a Quechan, and to be able to point visitors to the river instead of warning them to stay away for their own safety.

As I let the river lead me back into town, I think I can see a new kind of southwestern identity taking shape from the ruins of the old. On the overlook to my right is Fort Yuma. On the bank below, a Quechan man throws a stick into the river for his dog. As I coast beneath the Ocean-to-Ocean Bridge, I listen to the flow of people in both directions. I pass Gateway Park, where clusters of kids chit-chat on the beach, soaking up the sun. American coots dodge my paddleboard and hustle upriver, tooting their tiny tin horns. Up ahead the river turns a corner. A finger of wind riffles the surface, and cottonwoods shimmer in the breeze.

LESLIE JAMISON

The Empathy Exams

A Medical Actor Writes Her Own Script

FROM *The Believer*

DISCUSSED: Inexplicable Seizures, An Ailing Plastic Baby, Teenagers in Ponchos, An Endless Supply of Mints, Another Word for Burning, Crippled Rabbits in Love, The Sad Half-Life of Arguments, A Kid's Drawing of God, Praying in the Nook, Major Personality Clusters, fMRI Scans, Adam Smith, Pet Fears, Impulse's Dowdier Cousin, A Broken Arrow, A Bottle of Rain

My job title is Medical Actor, which means I play sick. I get paid by the hour. Medical students guess my maladies. I'm called a Standardized Patient, which means I act toward the norms of my disorders. I'm standardized-lingo SP for short. I'm fluent in the symptoms of preeclampsia and asthma and appendicitis. I play a mom whose baby has blue lips.

Medical acting works like this: You get a script and a paper gown. You get $13.50 an hour. Our scripts are 10 to 12 pages long. They outline what's wrong with us—not just what hurts but how to express it. They tell us how much to give away, and when. We are supposed to unfurl the answers according to specific protocols. The scripts dig deep into our fictive lives: the ages of our children and the diseases of our parents, the names of our husbands' real estate and graphic design firms, the amount of weight we've lost in the past year, the amount of alcohol we drink each week.

My specialty case is Stephanie Phillips, a 23-year-old who suffers from something called conversion disorder. She is grieving the death of her brother, and her grief has sublimated into seizures. Her disorder is news to me. I didn't know you could have a seizure from sadness. She's not supposed to know either. She's not supposed to think the seizures have anything to do with what she's lost.

Stephanie Phillips

PSYCHIATRY

SP TRAINING MATERIALS

Case Summary: You are a 23-year-old female patient experiencing seizures with no identifiable neurological origin. You can't remember your seizures but are told you froth at the mouth and yell obscenities. You can usually feel a seizure coming before it arrives. The seizures began two years ago, shortly after your older brother drowned in the river just south of the Bennington Avenue Bridge. He was swimming drunk after a football tailgate. You and he worked at the same minigolf course. These days you don't work at all. These days you don't do much. You're afraid of having a seizure in public. No doctor has been able to help you. Your brother's name was Will.

Medication History: You are not taking any medications. You've never taken antidepressants. You've never thought you needed them.

Medical History: Your health has never caused you any trouble. You've never had anything worse than a broken arm. Will was there when it was broken. He was the one who called for the paramedics and kept you calm until they came.

Our simulated exams take place in three suites of purpose-built rooms. Each room is fitted with an examination table and a surveillance camera. We test second- and third-year medical students in topical rotations: pediatrics, surgery, psychiatry. On any given day of exams, each student must go through "encounters"—their technical title—with three or four actors playing different cases.

A student might have to palpate a woman's 10-on-a-scale-of-10 pain in her lower abdomen, then sit across from a delusional young lawyer and tell him that when he feels a writhing mass of worms in his small intestine, the feeling is probably coming from somewhere else. Then this med student might arrive in my room, stay straight-faced, and tell me that I might go into premature labor to deliver the pillow strapped to my belly, or nod solemnly as I express concern about my ailing plastic baby: "He's just so quiet."

Once the 15-minute encounter has finished, the medical student leaves the room and I fill out an evaluation of his/her performance. The first part is a checklist: which crucial pieces of information did he/she manage to elicit? Which ones did he/she leave uncovered? The second part of the evaluation covers affect. Checklist item 31 is generally acknowledged as the most important category: "Voiced empathy for my situation/problem." We are instructed about the importance of this first word, *voiced.* It's not enough for someone to have a sympathetic manner or use a caring tone of voice. The students have to say the right words to get credit for compassion.

We SPs are given our own suite for preparation and decompression. We gather in clusters: old men in crinkling blue robes, MFA graduates in boots too cool for our paper gowns, local teenagers in ponchos and sweatpants. We help each other strap pillows around our waists. We hand off infant dolls. Little pneumonic Baby Doug, swaddled in a cheap cotton blanket, is passed from girl to girl like a relay baton. Our ranks are full of community-theater actors and undergrad drama majors seeking stages, high school kids earning booze money, retired folks with spare time. I am a writer, which is to say, I'm trying not to be broke.

We play a demographic menagerie: young jocks with ACL injuries and business executives nursing coke habits. STD Grandma has just cheated on her husband of 40 years and has a case of gonorrhea to show for it. She hides behind her shame like a veil, and her med student is supposed to part the curtain. If he's asking the right questions, she'll have a simulated crying breakdown halfway through the encounter.

Blackout Buddy gets makeup: a gash on his chin, a black eye, and bruises smudged in green eye shadow along his cheekbone. He's been in a minor car crash he can't remember. Before the encounter the actor splashes booze on his body like cologne. He's supposed to let the particulars of his alcoholism glimmer through, very "unplanned," bits of a secret he's done his best to keep guarded.

Our scripts are studded with moments of flourish: Pregnant Lila's husband is a yacht captain sailing overseas in Croatia. Appendicitis Angela has a dead guitarist uncle whose tour bus was hit by a tornado. Many of our extended family members have died violent, midwestern deaths: mauled in tractor- or grain-elevator accidents, hit by drunk drivers on the way home from Hy-Vee grocery stores, felled by a Big Ten tailgate—or, like my brother Will, by the aftermath of its debauchery.

Between encounters we are given water, fruit, granola bars, and an endless supply of mints. We aren't supposed to exhaust the students with our bad breath and growling stomachs, the side effects of our actual bodies.

Some med students get nervous during our encounters. It's like an awkward date, except half of them are wearing platinum wedding bands. I want to tell them I'm more than just an unmarried woman faking seizures for pocket money. *I do things!* I want to tell them. *I'm probably going to write about this in a book someday!* We make small talk about the rural Iowa farm town I'm supposed to be from. We each understand the other is inventing this small talk and we agree to respond to each other's inventions as genuine exposures of personality. We're holding the fiction between us like a jump rope.

One time a student forgets we are pretending and starts asking detailed questions about my fake hometown—which, as it happens, if he's being honest, is his *real* hometown—and his questions lie beyond the purview of my script, beyond what I can answer, because in truth I don't know much about the person I'm supposed to be or the place I'm supposed to be from. He's forgotten our contract. I bullshit harder, more heartily. "That park in Muscatine!" I say, slapping my knee like a grandpa. "I used to sled there as a kid."

Other students are all business. They rattle through the clinical checklist for depression like a list of things they need to get at the grocery store: "sleep disturbances, changes in appetite, decreased concentration." Some of them get irritated when I obey my script and refuse to make eye contact. I'm supposed to stay swaddled and numb. These irritated students take my averted eyes as a challenge. They never stop seeking my gaze. Wrestling me into eye contact is the way they maintain power, forcing me to acknowledge their requisite display of care.

I grow accustomed to comments that feel aggressive in their

formulaic insistence: *That must really be hard* [to have a dying baby], *That must really be hard* [to be afraid you'll have another seizure in the middle of the grocery store], *That must really be hard* [to carry in your uterus the bacterial evidence of cheating on your husband]. Why not say, *I couldn't even imagine*?

Other students seem to understand that empathy is always perched precariously between gift and invasion. They won't even press the stethoscope to my skin without asking if it's okay. They need permission. They don't want to presume. Their stuttering unwittingly honors my privacy: "Can I . . . could I . . . would you mind if I—listened to your heart?" "No," I tell them. "I don't mind." Not minding is my job. Their humility is a kind of compassion in its own right. Humility means they ask questions, and questions mean they get answers, and answers mean they get points on the checklist: a point for finding out my mother takes Wellbutrin, a point for getting me to admit I've spent the last two years cutting myself, a point for finding out my father died in a grain elevator when I was two—for realizing that a root system of loss stretches radial and rhizomatic under the entire territory of my life.

In this sense, empathy isn't measured just by checklist item 31—"Voiced empathy for my situation/problem"—but by every item that gauges how thoroughly my experience has been imagined. Empathy isn't just remembering to say *That must really be hard,* it's figuring out how to bring difficulty into the light so it can be seen at all. Empathy isn't just listening, it's asking the questions whose answers need to be listened to. Empathy requires inquiry as much as imagination. Empathy requires knowing you know nothing. Empathy means acknowledging a horizon of context that extends perpetually beyond what you can see: an old woman's gonorrhea is connected to her guilt is connected to her marriage is connected to her children is connected to the days when she was a child. All this is connected to her domestically stifled mother, in turn, and to her parents' unbroken marriage; maybe everything traces its roots to her very first period, how it shamed and thrilled her.

Empathy means realizing no trauma has discrete edges. Trauma bleeds. Out of wounds and across boundaries. Sadness becomes a seizure. Empathy demands another kind of porousness in response. My Stephanie script is 12 pages long. I think mainly about what it doesn't say.

Empathy comes from the Greek *empatheia*—*em* ("into") and *pathos* ("feeling")—a penetration, a kind of travel. It suggests you enter another person's pain as you'd enter another country, through immigration and customs, border-crossing by way of query: *What grows where you are? What are the laws? What animals graze there?*

I've thought about Stephanie Phillips's seizures in terms of possession and privacy—that converting her sadness away from direct articulation is a way to keep it hers. Her refusal to make eye contact, her unwillingness to explicate her inner life, the very fact that she becomes unconscious during her own expressions of grief and doesn't remember them afterward—all of these might be ways of keeping her loss protected and pristine, unviolated by the sympathy of others.

"What do you call out during seizures?" one student asks.

"I don't know," I say, and want to add, *but I mean all of it.*

I know that saying this would be against the rules. I'm playing a girl who keeps her sadness so subterranean she can't even see it herself. I can't give it away so easily.

Leslie Jamison

OB-GYN

SP TRAINING MATERIALS

Case Summary: You are a 25-year-old female seeking termination of your pregnancy. You have never been pregnant before. You are five and a half weeks but have not experienced any bloating or cramping. You have experienced some fluctuations in mood but have been unable to determine whether these are due to being pregnant or knowing you are pregnant. You are not visibly upset about your pregnancy. Invisibly, you are not sure.

Medication History: You are not taking any medications. This is why you got pregnant.

Medical History: You've had several surgeries in the past but you don't mention them to your doctor because they don't seem relevant. You are about to have another surgery to correct your tachycardia, the excessive and irregular beating of your heart. Your mother has made you promise to mention this upcoming surgery in your termination consultation, even though you don't feel like discussing it. She wants the doctor to know about your heart condition in case it affects the way he ends your pregnancy, or the way he keeps you sedated while he does it.

I could tell you I got an abortion one February or heart surgery that March—like they were separate cases, unrelated scripts—but neither one of these accounts would be complete without the other. A single month knitted them together; two mornings I woke up on an empty stomach and slid into a paper gown. One operation depended on a tiny vacuum, the other on a catheter that would ablate the tissue of my heart. *Ablate?* I asked the doctors. They explained that meant "burn."

One procedure made me bleed and the other was nearly bloodless; one was my choice and the other wasn't; both made me feel—at once—the incredible frailty and capacity of my own body; both came in a bleak winter; both left me prostrate under the hands of men, and dependent on the care of a man I was just beginning to love.

Dave and I first kissed in a Maryland basement at three in the morning on our way to Newport News to canvass for Obama in 2008. We canvassed for an organizing union called Unite Here. *Unite Here!* Years later that poster hung above our bed. That first fall we walked along Connecticut beaches strewn with broken clamshells. We held hands against salt winds. We went to a hotel for the weekend and put so much bubble bath in our tub that the bubbles ran all over the floor. We took pictures of that. We took pictures of everything. We walked across Williamsburg in the rain to see a concert. We were writers in love. My boss used to imagine us curling up at night and taking inventories of each other's hearts. *How did it make you feel to see that injured pigeon in the street today?*, etc. And it's true: we once talked about seeing two crippled bunnies trying to mate on a patchy lawn—how sad it was, and moving.

We'd been in love about two months when I got pregnant. I saw the cross on the stick and called Dave and we wandered college quads in the bitter cold and talked about what we were going to do. I thought of the little fetus bundled inside my jacket with me and wondered—honestly *wondered*—if I felt attached to it yet. I wasn't sure. I remember not knowing what to say. I remember wanting a drink. I remember wanting Dave to be inside the choice with me but also feeling possessive of what was happening. I needed him to understand he would never live this choice like I was going to live it. This was the double blade of how I felt about anything that hurt: I wanted someone else to feel it with me, and also I wanted it entirely for myself.

We scheduled the abortion for a Friday and I found myself facing a week of ordinary days until it happened. I realized I was supposed to keep doing ordinary things. One afternoon I holed up in the library and read a pregnancy memoir. The author described a pulsing fist of fear and loneliness inside her—a fist she'd carried her whole life, had numbed with drinking and sex—and explained how her pregnancy had replaced this fist with the tiny bud of her fetus, a moving life.

I sent Dave a text. I wanted to tell him about the fist of fear, the baby heart, how sad it felt to read about a woman changed by pregnancy when I knew I wouldn't be changed by mine—or at least not like she'd been. I didn't hear anything back for hours. This bothered me. I felt guilt that I didn't feel more about the abortion; I felt pissed off at Dave for being elsewhere, for choosing not to do the tiniest thing when I was going to do the rest of it.

I felt the weight of expectation on every moment—the sense that the end of this pregnancy was something I *should* feel sad about, the lurking fear that I never felt sad about what I was supposed to feel sad about, the knowledge that I'd gone through several funerals dry-eyed, the hunch that I had a parched interior life activated only by the need for constant affirmation, nothing more. I wanted Dave to guess what I needed at precisely the same time as I needed it. I wanted him to imagine how much small signals of his presence might mean.

That night we roasted vegetables and ate them at my kitchen table. Weeks before, I'd covered that table with citrus fruits and fed our friends pills made from berries that made everything sweet: grapefruit tasted like candy, beer like chocolate, Shiraz like Manischewitz—everything, actually, tasted a little like Manischewitz. Which is to say, that kitchen held the ghosts of countless days that felt easier than the one we were living now. We drank wine and I think—I know—I drank a lot. It sickened me to think I was doing something harmful to the fetus, because that meant thinking of the fetus as harmable, which made it feel more alive, which made me feel more selfish, woozy with cheap Cabernet and spoiling for a fight.

Feeling Dave's distance that day had made me realize how

much I needed to feel he was as close to this pregnancy as I was —an impossible asymptote. But I thought he could at least bridge the gap between our days and bodies with a text. I told him so. Actually, I probably sulked, waited for him to ask, and then told him so. *Guessing your feelings is like charming a cobra with a stethoscope,* a boyfriend told me once. Meaning what? Meaning a couple things, I think—that pain turned me venomous, that diagnosing me required a specialized kind of enchantment, that I flaunted feelings and withheld their origins at once.

Sitting with Dave in my attic living room, my cobra hood was spread. "I felt lonely today," I told him. "I wanted to hear from you."

I'd be lying if I wrote that I remember what he said. I don't. Which is the sad half-life of arguments—we usually remember our side better. I think he told me he'd been thinking of me all day, and couldn't I trust that? Why did I need proof?

Voiced empathy for my situation/problem. Why did I need proof? I just did.

He said to me, "I think you're making this up."

"This" meaning what? My anger? My anger at him? Memory fumbles.

I didn't know what I felt, I told him. Couldn't he just trust that I felt something, and that I'd wanted something from him? I needed his empathy not just to comprehend the emotions I was describing but to help me discover which emotions were actually there.

We were under a skylight under a moon. It was February beyond the glass. It was almost Valentine's Day. I was curled into a cheap futon with crumbs in its creases, a piece of furniture that made me feel like I was still in college. This abortion was something adult. I didn't feel like an adult inside of it.

I heard "making this up" as an accusation that I was inventing emotions I didn't have, but I think he was suggesting I'd mistranslated emotions that were already there—attaching long-standing feelings of need and insecurity to the particular event of this abortion; exaggerating what I felt in order to manipulate him into feeling bad. This accusation hurt not because it was entirely wrong but because it was partially right, and because it was leveled with such coldness. He was speaking something truthful about me in order to defend himself, not to make me feel better.

But there was truth behind it. He understood my pain as something actual and constructed at once. He got that it was necessarily both—that my feelings were also made of the way I spoke them. When he told me I was making things up, he didn't mean I wasn't feeling anything. He meant that feeling something was never simply a state of submission but always, also, a process of construction. I see all this, looking back.

I also see that he could have been gentler with me. We could have been gentler with each other.

We went to Planned Parenthood on a freezing morning. We rummaged through a bin of free kids' books while I waited for my name to get called. Who knows why these children's books were there? Meant for kids waiting during their mothers' appointments, maybe. But it felt like perversity that Friday morning, during the weekly time slot for abortions. We found a book called *Alexander,* about a boy who confesses all his misdeeds to his father by blaming them on an imaginary red-and-green-striped horse. "Alexander was a pretty bad horse today." Whatever we can't hold, we hang on a hook that will hold it. The book belonged to a guy named Michael from Branford. I wondered why Michael had come to Planned Parenthood, and why he'd left that book behind.

There are things I'd like to tell the version of myself who sat in the Planned Parenthood counseling room, the woman who studiously practiced cheerful unconcern. I would tell her she is going through something large and she shouldn't be afraid to confess its size, shouldn't be afraid she's "making too big a deal of it." She shouldn't be afraid of not feeling enough, because the feelings will keep coming—different ones—for years. I would tell her that commonality doesn't inoculate against hurt. The fact of all those women in the waiting room, doing the same thing I was doing, didn't make it any easier.

I would tell myself: Maybe your prior surgeries don't matter here, but maybe they do. Your broken jaw and your broken nose don't have anything to do with your pregnancy except that they were both times you got broken into. Getting each one fixed meant getting broken into again. Getting your heart fixed will be another burglary, nothing taken except everything that gets burned away. Maybe every time you get into a paper gown you summon the ghosts of all the other times you got into a paper gown; maybe every time you slip down into that anesthetized dark it's the same dark you slipped into before. Maybe it's been waiting for you the whole time.

Stephanie Phillips

PSYCHIATRY

SP TRAINING MATERIALS (CONT.)

Opening Line: "I'm having these seizures and no one knows why."

Physical Presentation and Tone: You are wearing jeans and a sweatshirt, preferably stained or rumpled. You aren't someone who puts much effort into your personal appearance. At some point during the encounter, you might mention that you don't bother dressing nicely anymore, because you rarely leave the house. It is essential that you avoid eye contact and keep your voice free of emotion during the encounter.

One of the hardest parts of playing Stephanie Phillips is nailing her affect—*la belle indifference,* a manner defined as the "air of unconcern displayed by some patients toward their physical symptoms." It is a common sign of conversion disorder, a front of indifference hiding "physical symptoms [that] may relieve anxiety and result in secondary gains in the form of sympathy and attention given by others." *La belle indifference*—outsourcing emotional content to physical expression—is a way of inviting empathy without asking for it. In this way encounters with Stephanie present a sort of empathy limit case: the clinician must excavate a sadness the patient hasn't identified, must imagine deeply into a pain Stephanie can't fully experience herself.

For other cases we are supposed to wear our anguish more openly—like a terrible, seething garment. My first time playing Appendicitis Angela, I'm told I manage "just the right amount of pain." I'm moaning in a fetal position and apparently doing it right. The doctors know how to respond. "I am sorry to hear that you are experiencing an excruciating pain in your abdomen," one says. "It must be uncomfortable."

Part of me has always craved a pain so visible—so irrefutable and physically inescapable—that everyone would have to notice.

But my sadness about the abortion was never a convulsion. There was never a scene. No frothing at the mouth. I was almost relieved, three days after the procedure, when I started to hurt. It was worst at night, the cramping. But at least I knew what I felt. I wouldn't have to figure out how to explain. Like Stephanie, who didn't talk about her grief, because her seizures were already pronouncing it —slantwise, in a private language, but still—granting it substance and choreography.

Stephanie Phillips

PSYCHIATRY

SP TRAINING MATERIALS (CONT.)

Encounter Dynamics: You don't reveal personal details until prompted. You say you wouldn't call yourself happy. You say you wouldn't call yourself unhappy. You get sad some nights about your brother. You don't say so. You don't say you have a turtle who might outlive you, and a pair of green sneakers from your gig at the mini-golf course. You don't say you have a lot of memories of stacking putters. You say you have another brother, if asked, but you don't say he's not Will, because that's obvious—even if the truth of it still strikes you sometimes, hard. You are not sure these things matter. They are just facts. They are facts like the fact of dried spittle on your cheeks when you wake up on the couch and can't remember telling your mother to fuck herself. Fuck you is also what your arm says when it jerks so hard it might break into pieces. Fuck you fuck you fuck you until your jaw locks and nothing comes.

You live in a world underneath the words you are saying in this clean white room, It's okay I'm okay I feel sad I guess. You are blind in this other world. It's dark. Your seizures are how you move through it —thrashing and fumbling—feeling for what its walls are made of.

Your body wasn't anything special until it rebelled. Maybe you thought your thighs were fat or else you didn't, yet; maybe you had best friends who whispered secrets to you during sleepovers; maybe you had lots of boyfriends or else you were still waiting for the first one; maybe you liked unicorns when you were young or maybe you preferred regular horses. I imagine you in every possible direction, and then I cover my tracks and imagine you all over again. Sometimes I can't stand how much of you I don't know.

*

I hadn't planned to get heart surgery right after an abortion. I hadn't planned to get heart surgery at all. It came as a surprise that there was anything wrong. My pulse had been showing up high at the doctor's office. I was given a Holter monitor—a small plastic box to wear around my neck for 24 hours, attached by sensors to my chest—that showed the doctors my heart wasn't beating right. The doctors diagnosed me with SVT—superventricular tachycardia—and said they thought there was an extra electrical node sending out extra signals—*beat, beat, beat*—when it wasn't supposed to.

They explained how to fix it: they'd make two slits in my skin, above my hips, and thread catheter wires all the way up to my heart. They would ablate bits of tissue until they managed to get rid of my tiny rogue beatbox.

My primary cardiologist was a small woman who moved quickly through the offices and hallways of her world. Let's call her Dr. M. She spoke in a curt voice, always. The problem was never that her curtness meant anything—never that I took it personally—but rather that it meant nothing, that it wasn't personal at all.

My mother insisted I call Dr. M to tell her I was having an abortion. What if there was something I needed to tell the doctors before they performed it? That was the reasoning. I put off the call until I couldn't put it off any longer. The thought of telling a near stranger that I was having an abortion—over the phone, without being asked—seemed mortifying. It was like I'd be peeling off the bandage on a wound she hadn't asked to see.

When I finally got her on the phone, she sounded harried and impatient. I told her quickly. Her voice was cold: "And what do you want to know from me?"

I went blank. I hadn't known I'd wanted her to say, *I'm sorry to hear that,* until she didn't say it. But I had. I'd wanted her to say something. I started crying. I felt like a child. I felt like an idiot. Why was I crying now, when I hadn't cried before—not when I found out, not when I told Dave, not when I made the appointment or went to it?

"Well?" she asked.

I finally remembered my question: did the abortion doctor need to know anything about my tachycardia?

"No," she said. "Is that it?" Her voice was so incredibly blunt. I could hear only one thing in it: *Why are you making a fuss?* That was it. I felt simultaneously like I didn't feel enough and like I was making a big deal out of nothing—that maybe I was making a big deal out of nothing *because* I didn't feel enough, that my tears with Dr. M were runoff from the other parts of the abortion I wasn't crying about. I had an insecurity that didn't know how to express itself, that could attach itself to tears or else to their absence. *Alexander was a pretty bad horse today.* When of course the horse wasn't the problem. Dr. M became a villain because my story didn't have one. Mine was the kind of pain that comes without a perpetrator. Everything was happening because of my body or because of a choice I'd made. I needed something from the world I didn't know how to ask for. I needed people—Dave, a doctor, anyone—to deliver my feelings back to me in a form that was legible. Which is a superlative kind of empathy to seek, or to supply: an empathy that rearticulates more clearly what it's shown.

A month later Dr. M bent over the operating table and apologized. "I'm sorry for my tone on the phone," she said. "I didn't understand what you were asking." It was an apology whose logic I didn't entirely understand. It had been prompted. At some point my mother had called her to discuss my upcoming procedure—and had mentioned how upset I'd been by our phone conversation.

Now I was lying on my back in a hospital gown. I was woozy from the early stages of my anesthesia. I felt like crying all over again, at the memory of how powerless I'd been on the phone—powerless because I had needed so much from her, a stranger—and how powerless I was now, lying flat on my back and waiting for a team of doctors to burn away the tissue of my heart. I wanted to tell her I didn't accept her apology. I wanted to tell her she didn't have the right to apologize—not here, not while I was lying naked under a paper gown, not when I was about to get cut open again. I wanted to deny her the right to feel better because she'd said she was sorry.

Mainly I wanted the anesthesia to carry me away from everything I felt and everything my body was about to feel. In a moment it did.

I always fight the impulse to ask the med students for pills during our encounters. It seems natural. Wouldn't Baby Doug's mom want an Ativan? Wouldn't Appendicitis Amy want some Vicodin, or whatever they give you for a 10 on the pain scale? Wouldn't Stephanie Phillips be a little more excited about a new diet of Valium? I keep thinking I'll communicate my pain most effectively by expressing my desire for the things that might dissolve it. Which is to say, if I were Stephanie Phillips, I'd be excited about my Ativan. But I'm not. And being an SP isn't about projection; it's about inhabitance. I can't go off-script. These encounters aren't about dissolving pain, anyway, but rather seeing it more clearly. The healing part is always a hypothetical horizon we never reach.

During my winter of ministrations, I found myself constantly in the hands of doctors. It began with that first nameless man who gave me an abortion the same morning he gave 20 other women their abortions. *Gave.* It's a funny word we use, as if it were a present. Once the procedure was done, I was wheeled into a dim room where a man with a long white beard gave me a cup of orange juice. He was like a kid's drawing of God. I remember resenting how he wouldn't give me any pain pills until I'd eaten a handful of crackers, but he was kind. His resistance was a kind of care. I felt that. He was looking out for me.

Dr. G was the doctor who performed my heart operation. He controlled the catheters from a remote computer. It looked like a spaceship flight cabin. He had a nimble voice and lanky arms and bushy white hair. I liked him. He was a straight talker. He came into my hospital room the day after my operation and explained why the procedure hadn't worked: they'd burned and burned, but they hadn't burned the right patch. They'd even cut through my arterial wall to keep looking. But then they'd stopped. Ablating more tissue risked dismantling my circuitry entirely.

Dr. G said I could get the procedure again. I could authorize them to ablate more aggressively. The risk was that I'd come out of surgery with a pacemaker. He was very calm when he said this. He pointed at my chest: "On someone thin," he said, "you'd be able to see the outlines of the box quite clearly."

I pictured waking up from general anesthesia to find a metal box above my ribs. I remember being struck by how the doctor had anticipated a question about the pacemaker I hadn't yet discovered in myself: How easily would I be able to forget it was there? I remember feeling grateful for the calmness in his voice and not offended by it. It didn't register as callousness. Why? Maybe it was just because he was a man. I didn't need him to be my mother—even for a day—I only needed him to know what he was doing. But I think it was something more.

Instead of identifying with my panic—inhabiting my horror at the prospect of a pacemaker—he was helping me understand that even this, the barnacle of a false heart, would be okay. His calmness didn't make me feel abandoned, it made me feel secure. It offered assurance rather than empathy, or maybe assurance was evidence of empathy, insofar as he understood that assurance, not identification, was what I needed most.

Empathy is a kind of care but it's not the only kind of care, and it's not always enough. I want to think that's what Dr. G was thinking. I needed to look at him and see the opposite of my fear, not its echo.

Every time I met with Dr. M, she began our encounters with a few perfunctory questions about my life—"What are you working on these days?"—and when she left the room to let me dress, I could hear her voice speaking into a tape recorder in the hallway: "Patient is a graduate student in English at Yale. Patient is writing a dissertation on addiction. Patient spent two years living in Iowa. Patient is working on a collection of essays." And then, without fail, at the next appointment, fresh from listening to her old tape, she bulleted a few questions: "How were those two years in Iowa? How's that collection of essays?"

It was a strange intimacy, almost embarrassing, to feel the mechanics of her method so palpably between us: "Engage the patient, record the details, repeat." I was sketched into CliffsNotes. I hated seeing the puppet strings; they felt unseemly—and without kindness in her voice, the mechanics meant nothing. They pretended we knew each other rather than acknowledging that we didn't. It's a tension intrinsic to the surgeon-patient relationship: it's more invasive than anything but not intimate at all.

Now I can imagine another kind of tape—a more naked, stuttering tape; a tape that keeps correcting itself, that messes up its dance steps:

Patient is here for ~~an abortion~~ ~~a surgery to burn the bad parts of her heart~~ a medication to fix her heart because the surgery failed. Patient is staying in the hospital for ~~one night~~ ~~three nights~~ five nights until we get this medication right. Patient ~~wonders if people can bring her booze in the hospital~~ likes to eat graham crackers from the nurses' station. Patient cannot be released until she runs on a treadmill and her heart prints a clean rhythm. Patient recently got an abortion but we don't understand why she wanted us to know that. Patient didn't ~~think she~~ hurt at first but then she did. Patient ~~failed to use protection and~~ failed to provide an adequate account of why she didn't use protection. ~~Patient had a lot of feelings. Partner of patient had the feeling she was making up a lot of feelings.~~ Partner of patient is supportive. Partner of patient is spotted in patient's hospital bed, repeatedly. Partner of patient is caught kissing patient. Partner of patient is charming.

Patient is ~~angry~~ ~~disappointed~~ angry her procedure failed. Patient does not want to be on medication. Patient wants to know if she can drink alcohol on this medication. She wants to know how much. She wants to know ~~if two bottles of wine a night is too many~~ if she can get away with a couple glasses. Patient does not want to get another procedure if it means risking a pacemaker. Patient wants everyone to understand that this surgery ~~is~~ isn't a big deal; wants everyone to understand she is stupid for crying when everyone else on the ward is sicker than she is; wants everyone to understand her abortion is ~~also about~~ definitely not about the children her ex-boyfriends have had since she broke up with them. Patient wants everyone to understand ~~it wasn't a choice~~ it would have been easier if it hadn't been a choice. Patient understands it was her choice to drink while she was pregnant. She understands it was her choice to go to a bar with a little plastic box hanging from her neck and get so drunk she messed up her heart graph. Patient is patients, plural, which is to say she is multiple—mostly grateful but sometimes surly, sometimes full of self-pity. Patient ~~already understands~~ is trying hard to understand she needs to listen up if she wants to hear how everyone is caring for her.

Three men waited for me in the hospital during my surgery: my brother and my father and Dave. They sat in the lounge making awkward conversation, and then in the cafeteria making awkward conversation, and then—I'm not sure where they sat, actually, or in what order, because I wasn't there. But I do know that while they were sitting in the cafeteria a doctor came to find them and told them that the surgeons were going to tear through part of my arterial wall—these were the words they used, Dave said, *tear through*—and try burning some patches of tissue on the other side.

At this point, Dave told me later, he went to the hospital chapel and prayed I wouldn't die. He prayed in the nook made by the propped-open door, because he didn't want to be seen.

It wasn't likely I would die. He didn't know that then. Prayer isn't about likelihood anyway, it's about desire—loving someone enough to get on your knees and ask for her to be saved. When he cried in that chapel, it wasn't empathy—it was something else. His kneeling wasn't a way to feel my pain but to request that it end.

I learned to rate Dave on how well he empathized with me. I was constantly poised above an invisible checklist item 31. I wanted him to hurt whenever I hurt, to feel as much as I felt. But it's exhausting to keep tabs on how much someone is feeling for you. It can make you forget that they feel too.

I used to believe that hurting would make you more alive to the hurting of others. I used to believe in feeling bad because somebody else did. Now I'm not so sure of either. I know that being in the hospital made me selfish. Getting surgeries made me think mainly about whether I'd have to get another one. When bad things happened to other people, I imagined them happening to me. I didn't know if this was empathy or theft.

For example: one September my brother woke up in a hotel room in Sweden and couldn't move half his face. He was diagnosed with something called Bell's palsy. No one really understands why it happens or how to make it better. The doctors gave him a steroid called prednisone that made him sick. He threw up most days around twilight. He sent us a photo. It was grainy. He looked lonely. His face slumped. His pupil glistened in the flash, bright with the gel he had to put on his eye to keep it from drying out. He couldn't blink.

I found myself obsessed with his condition. I tried to imagine what it was like to move through the world with an unfamiliar face. I thought about what it would be like to wake up in the morning, in the groggy space where you've managed to forget things, to forget your whole life, and then snapping to, realizing: Yes, this is how things are. Checking the mirror: still there. I tried to imagine how you'd feel a little crushed each time, coming out of dreams to another day of being awake with a face not quite your own.

I spent large portions of each day—pointless, fruitless spans of time—imagining how I would feel if my face was paralyzed too.

I stole my brother's trauma and projected it onto myself like a magic-lantern pattern of light. I obsessed, and told myself this obsession was empathy. But it wasn't, quite. It was more like *in*pathy. I wasn't expatriating myself into another life so much as importing its problems into my own.

Dave doesn't believe in feeling bad just because someone else does. This isn't his notion of support. He believes in listening, and asking questions, and steering clear of assumptions. He thinks imagining someone else's pain with too much surety can be as damaging as failing to imagine it. He believes in humility. He believes in staying strong enough to stick around. He stayed with me in the hospital, five nights in those crisp white beds, and he lay down with my monitor wires, colored strands carrying the electrical signature of my heart to a small box I held in my hands. I remember lying tangled with him, how much it meant —that he was willing to lie down in the mess of wires, to stay there with me.

In order to help the med students empathize better with us, we have to empathize with them. I try to think about what makes them fall short of what they're asked to do—what nervousness or squeamishness or callousness—and how to speak to their sore spots without bruising them: the one so stiff he shook my hand like we'd just made a business deal; the chipper one so eager to befriend me she hadn't washed her hands at all.

One day we have a sheet cake delivered for my supervisor's birthday—dry white cake with ripples of strawberry jelly between its layers—and we sit around our conference table eating her cake with plastic forks while she doesn't eat anything at all. She tells us what kind of syntax we should use when we tell the students about bettering their empathy. We're supposed to use the "When you . . . I felt" frame. *When you forgot to wash your hands, I felt protective of my body. When you told me 11 wasn't on the pain scale, I felt dismissed.* For the good parts also: *When you asked me questions about Will, I felt like you really cared about my loss.*

A 1983 study titled "The Structure of Empathy" found a correlation between empathy and four major personality clusters: sensitivity, nonconformity, even-temperedness, and social self-confidence.

I like the word *structure.* It suggests empathy is an edifice we build like a home or office—with architecture and design, scaffolding and electricity. The Chinese character for *listen* is built of many parts: the characters for *ear* and *eye,* the horizontal line that signifies undivided attention, the swoop and teardrops of *heart.*

Rating high for the study's "sensitivity" cluster feels intuitive. It means agreeing with statements like "I have at one time or another tried my hand at writing poetry," or "I have seen some things so sad they almost made me feel like crying," and *dis*agreeing with statements like "I really don't care whether people like me or dislike me." This last one seems to suggest that empathy might be, at root, a barter, a bid for others' affection: *I care about your pain* is another way to say *I care if you like me.* We care in order to be cared for. We care because we are porous. The feelings of others matter, they are *like* matter: they carry weight, exert gravitational pull.

It's the last cluster, social self-confidence, that I don't understand as well. I've always treasured empathy as the particular privilege of the invisible, the observers who are shy precisely *because* they sense so much—because it is overwhelming to say even a single word when you're sensitive to every last flicker of nuance in the room. "The relationship between social self-confidence and empathy is the most difficult to understand," the study admits. But its explanation makes sense: social confidence is a prerequisite but not a guarantee; it can "give a person the courage to enter the interpersonal world and practice empathetic skills." We should empathize from courage, is the point—and it makes me think about how much of my empathy comes from fear. I'm afraid other people's problems will happen to me, or else I'm afraid other people will stop loving me if I don't adopt their problems as my own.

Jean Decety, a psychologist at the University of Chicago, uses fMRI scans to measure what happens when someone's brain responds to another person's pain. He shows test subjects images of painful situations (hand caught in scissors, foot under door) and compares these scans to what a brain looks like when its body is actually in pain. Decety has found that imagining the pain of others activates the same three areas (prefrontal cortex, anterior insula, anterior cingulate cortex) activated in the experience of pain itself. I feel heartened by that correspondence. But I also wonder what it's good for.

During the months of my brother's Bell's palsy, whenever I woke up in the morning and checked my face for a fallen cheek, a drooping eye, a collapsed smile, I wasn't ministering to anyone. I wasn't feeling toward my brother so much as I was feeling toward a version of myself—a self that didn't exist but theoretically shared his misfortune.

I wonder if my empathy has always been this, in every case: just a bout of hypothetical self-pity projected onto someone else. Is this ultimately just solipsism? Adam Smith confesses in his *Theory of Moral Sentiments:* "When we see a stroke aimed and just ready to fall upon the leg or arm of another person, we naturally shrink and draw back our own leg or our own arm."

We care about ourselves. Of course we do. Maybe some good comes from it. If I imagine myself fiercely into my brother's pain, I get some sense, perhaps, of what he might want or need, because I think, *I would want this. I would need this.* But it also seems like a fragile pretext, turning his misfortunes into an opportunity to indulge pet fears of my own devising.

I wonder which parts of my brain are lighting up when the med students ask me, "How does that make you feel?" Or which parts of their brains are glowing when I say, "The pain in my abdomen is a ten." My condition isn't real. I know this. They know this. I'm simply going through the motions. They're simply going through the motions. But motions can be more than rote. They don't just express feeling; they can give birth to it.

Empathy isn't just something that happens to us—a meteor shower of synapses firing across the brain—it's also a choice we make: to pay attention, to extend ourselves. It's made of exertion, that dowdier cousin of impulse. Sometimes we care for another because we know we should, or because it's asked for, but this doesn't make our caring hollow. The act of choosing simply means we've committed ourselves to a set of behaviors greater than the sum of our individual inclinations: *I will listen to his sadness, even when I'm deep in my own.* To say "going through the motions"—this isn't reduction so much as acknowledgment of the effort—the labor, the *motions,* the dance—of getting inside another person's state of heart or mind.

This confession of effort chafes against the notion that empathy should always arise unbidden, that *genuine* means the same thing as *unwilled,* that intentionality is the enemy of love. But I believe in intention and I believe in work. I believe in waking up in the middle of the night and packing our bags and leaving our worst selves for our better ones.

Leslie Jamison

OB-GYN

SP TRAINING MATERIALS (CONT.)

Opening Line: You don't need one. Everyone comes here for the same reason.

Physical Presentation and Tone: Wear loose pants. You have been told to wear loose pants. Keep your voice steady and articulate. You are about to spread your legs for a doctor who won't ever know your name. You know the drill, sort of. Act like you do.

Encounter Dynamics: Answer every question like you're clarifying a coffee order. Be courteous and nod vigorously. Make sure your heart stays on the other side of the white wall behind you. If the nurse asks you whether you are sure about getting the procedure, say yes without missing a beat. Say yes without a trace of doubt. Don't mention the way you felt when you first saw the pink cross on the stick—that sudden expansive joy at the possibility of a child, at your own capacity to have one. Don't mention this single moment of joy, because it might make it seem as if you aren't completely sure about what you're about to do. Don't mention this single moment of joy, because it might hurt. It will feel—more than anything else does—like the measure of what you're giving up. It maps the edges of your voluntary loss.

Instead, tell the nurse you weren't using birth control but wasn't that silly and now you are going to start.

If she asks what forms of birth control you have used in the past, say condoms. Suddenly every guy you've ever slept with is in the room with you. Ignore them. Ignore the memory of that first time—all that fumbling, and then pain—while Rod Stewart crooned "Broken Arrow" from a boom box on the dresser. "Who else is gonna bring you a broken arrow? Who else is gonna bring you a bottle of rain?"

Say you used condoms but don't think about all the times you didn't—in an Iowan graveyard, in a little car by a dark river—and definitely don't say why, how the risk made you feel close to those boys, how you courted the incredible gravity of what your bodies could do together.

If the nurse asks about your current partner, you should say, *We are very committed,* like you are defending yourself against some legal charge. If the nurse is listening closely, she should hear fear nestled like an egg inside your certainty.

If the nurse asks whether you drink, say yes to that too. Of course you do. Like it's no big deal. Your lifestyle habits include drinking to excess. You do this even when you know there is a fetus inside you. You do it to forget there is a fetus inside you; or to feel like maybe this is just a movie about a fetus being inside you.

The nurse will eventually ask, *How do you feel about getting the procedure?* Tell her you feel sad but you know it's the right choice, because this seems like the right thing to say, even though it's a lie. You feel mainly numb. You feel numb until your legs are in the stirrups. Then you hurt. Whatever anesthesia comes through the needle in your arm only sedates you. Days later you feel your body cramping in the night —a deep, hot, twisting pain—and you can only lie still and hope it passes, beg for sleep, drink for sleep, resent Dave for sleeping next to you. You can only watch your body bleed like an inscrutable, stubborn object—something harmed and cumbersome and not entirely yours. You leave your body and don't come back for a month. You come back angry.

You wake up from another round of anesthesia and they tell you all their burning didn't burn away the part of your heart that was broken. You come back and find you aren't alone. You weren't alone when you were cramping through the night and you're not alone now. Dave spends every night in the hospital. You want to tell him how disgusting your body feels: your unwashed skin and greasy hair. You want him to listen, for hours if necessary, and feel everything exactly as you feel it —your pair of hearts in such synchronized rhythm any monitor would show it; your pair of hearts playing two crippled bunnies doing whatever they can. There is no end to this fantasy of closeness. *Who else is gonna bring you a broken arrow?* You want him to break with you. You want him to hurt in a womb he doesn't have; you want him to admit he can't hurt that way. You want him to know how it feels in every one of your nerve endings: lying prone on the detergent sheets, lifting your shirt for one more cardiac resident, one more stranger, letting him attach his clips to the line of hooks under your breast, letting him print out your heart, once more, to see if its rhythm has calmed.

It all returns to this: you want him close to your damage. You want humility and presumption and whatever lies between, you want that too. You're tired of begging for it. You're tired of grading him on how well he gives it. You want to learn how to stop feeling sorry for yourself. You want to write an essay about the lesson. You throw away the checklist and let him climb into your hospital bed. You let him part the heart wires. You sleep. He sleeps. You wake, pulse feeling for another pulse, and there he is again.

BROOKE JARVIS

The Deepest Dig

FROM *The California Sunday Magazine*

ON THE NIGHTS before a dive, Cindy Lee Van Dover likes to stand on the deck of her research ship, looking down into the water the way an astronaut might look up at the stars.

She's preparing herself to do an extraordinary thing: climb into a tiny bubble of light and air and sink to the bottom of the ocean, leaving the sparkling waters of the surface a mile and a half above her.

She makes the trip in a three-person submersible called *Alvin,* famous for discovering the underwater hot springs known as hydrothermal vents and for exploring the wreckage of the *Titanic. Alvin* sinks for more than an hour. The view from its portholes moves through a spectrum of glowing greens and blues, eventually fading to pure black. The only break from the darkness comes when the sub drops through clusters of bioluminescence that look like stars in the Milky Way. They're the only way for Van Dover to tell, in the complete darkness and absence of acceleration, that she's sinking at all.

At last, as *Alvin* approaches the seafloor, the pilot turns on the external light. Van Dover peers hard, eager for her first glimpse of a strange land of underwater volcanoes and mountain ranges, of vast plains and smoking basalt spires.

It's the spires—the teetering chimneys that top hydrothermal vents—and their inhabitants that Van Dover has come to see. The animals that live on vents fascinate biologists like her because we understand so little about them. Scientists call them "alien" with only slight exaggeration: their most basic functions are unlike those of all other life on Earth, and astrobiologists study them to make better guesses about where to look for extraterrestrial life and what it might be like. "It allows us to see how life plays out on the next best thing to another planet," says David Grinspoon, chair of astrobiology at the Library of Congress.

But scientists aren't the only ones attracted to this strange world. The same vents that support colonies of undulating tubeworms, giant clams, eyeless shrimp, and hairy, tennis-ball-sized snails are also conduits for valuable metals fresh from the earth's interior. Vents form where seawater seeps into fissures in the earth's crust and reacts with the heat of magma, emerging transformed: acidic, boiling hot, and laden with chemicals and minerals. As the water cools, those minerals precipitate out, leaving behind concentrations of metals —gold, copper, nickel, and silver, as well as more esoteric minerals used in electronics —that make the richest mines on dry land look meager.

And where there's metal, there are miners, even at the bottom of the world.

As an industry, deep-sea mining is brand-new. The International Seabed Authority (ISA), which oversees all mining in international waters, was formed in 1994, but by 2011 it had issued only seven exploratory licenses. By the end of this year, it believes that number will jump to 26, with the first license for commercial mining expected as soon as 2016. Only one country, Papua New Guinea, has issued a permit for commercial-scale deep-sea mining in its own waters, though India, Japan, China, and South Korea also have projects in the early stages, and more than a dozen Pacific island nations, whose tiny populations and bureaucracies are dwarfed by their massive marine territories, are scrambling to figure out how to manage mining. Even for an industry that's seen plenty of false starts, says Michael Lodge, deputy to the secretary-general of the ISA, there is now "a hell of a lot of activity."

It's not news that we're looking beyond the usual places to find the things that power modern society: oil from the Arctic and bitumen from the tar sands, coltan mined by hand in the heart of the Congo, and natural gas fracked from beneath suburban backyards. We're even talking seriously about mining asteroids.

Still, there's something pause-worthy about mining the deep ocean. Literally unfathomable, the deep sea is still the most remote and least understood environment on Earth and perhaps the closest thing to a final frontier our beleaguered planet can claim. Jules Verne's Captain Nemo built the *Nautilus,* he declares, because the sea is the world's last refuge from humankind: "At thirty feet below its level, their reign ceases, their influence is quenched, and their power disappears."

Or maybe not. The first company to receive permission to mine the deep sea is called Nautilus Minerals.

Captain Nemo called the undersea world "the land of marvels." That's how Van Dover, growing up examining horseshoe crabs on the coast of New Jersey, saw it as well. From the moment she first read a scientific paper about hydrothermal vents, she became fascinated with the real-world denizens of the deep. She wrote to the paper's author, a scientist at the Smithsonian Institution, and he agreed to send her a small vial of vent-crab eggs to dissect and analyze. It arrived in the mail "as valuable to me as a moon rock," she says. Later, with the same scientist's help, she talked her way onto one of the first biological expeditions to hydrothermal vents. She was too junior to get a spot on a dive, so she busied herself studying dead squat lobsters that the pilots removed from the sub's exterior.

Van Dover eventually got her first dive in *Alvin* (she landed next to a bloom of crimson-plumed tubeworms known as the Rose Garden) and later earned her PhD in biological oceanography, but she realized that she craved more time on the seafloor than academia could offer her. So she took on the long, grueling process of training to be an *Alvin* pilot, becoming the 25th person and the first PhD-level scientist to do so. She was also the first —and is still the only —woman.

These days Van Dover is the chair of Duke University's Division of Marine Science and Conservation and the director of a lab that studies the population dynamics of organisms from hydrothermal-vent fields around the world. She's also something she never expected: a scientific consultant to Nautilus Minerals.

A Canadian company, Nautilus has obtained the mining exploration rights to nearly 200,000 square miles of territory throughout the Pacific, including in the Clarion-Clipperton Fracture Zone, a vast region of international waters southeast of Hawaii that's known to be rich in polymetallic nodules. Nautilus's first-in-the-world deep-sea mining permit gives it permission to dismantle a hydrothermal-vent site known as Solwara 1, located nearly a mile deep in Manus Basin in Papua New Guinea's Bismarck Sea.

Nautilus's plan for Solwara 1, which the company intends to begin mining in 2017, is to use two large robot excavators to remove chimneys and the first 160 feet of the seafloor. Other specially designed machines will grind this material to slurry and pipe it to the surface. There solids will be separated out and excess fluid (acidic and full of chemicals not ordinarily found in the upper ocean) will be pumped back down to the deep sea. The remaining material will be shipped to China, where a company called Tongling will extract gold, silver, and copper. Within the mined area, the vent structures and the animals that once lived on them—examples of which fill jars in Van Dover's lab—could disappear.

For Van Dover, that's a heartbreaking thought. "Because these are my babies, right?" she says. "This is stuff I've always held and revered." For years she and her fellow deep-sea scientists believed that the world they studied was far beyond the reach of industry. But as the most accessible land-based minerals are exhausted, those on the bottom of the sea are looking more like low-hanging fruit. "You have to go to more remote places," Lodge, of the ISA, says. "You have to go deeper."

In Tok Pisin, the lingua franca of Papua New Guinea (the country is home to more than 800 languages), *solwara* means "ocean"—*sol* plus *wara* equals "saltwater." The ocean is integral to life in New Ireland, the island province in whose waters Solwara 1 lies. People here farm coconuts, bananas, sago, and taro, plus *buai,* or betel nut, to sell at the market. Many don't use money for much beyond school fees, rides to town, and a few staple supplies. They raise pigs and hunt tree possums called cuscus. And they fish, sometimes far offshore.

Representatives from Nautilus have often visited New Ireland while developing plans for Solwara 1, but there's still a lot of uncertainty about what the mining project will mean. In Kontu, a village known for its shark-calling festival, I asked a woman named Helen Joel what questions she'd asked the representatives. "We do not ask questions because we do not know," she replied. During my three weeks in Papua New Guinea last winter, I repeatedly found people turning my own questions back on me: What *should* we think of the mine? What *does* it mean for the ocean? The one thing everyone seemed to know was that the mine would be the first, the very first, in the world.

January is the wet season in New Ireland, and the coastal road that edges the Bismarck Sea is unpaved. My rented Land Cruiser stalled repeatedly in deep water and was finally trapped between rain-swollen rivers in a tiny village whose name even my local companions didn't know. "Ugana," a man named Ray Wilfred told us. "Not Uganda. That's in Africa." Wilfred's neighbor, Ambrose Barais, a thickset man with a close-trimmed mustache and a thin, dark line tattooed on his right cheek, invited us to dinner with his wife and six children. I asked one of their sons his age, and he looked at me blankly before I remembered the right way to ask. "How many Christmases?" "Seventeen."

The rain poured without a break. Inside the house, by firelight, Barais's wife fried whole reef fish and boiled rice. One of the kids asked me what the metals from the seabed would be used for. It struck me how unlikely they were to end up back here, where a family's only metal possessions might be a few cooking pots and utensils, roofing, and sometimes a cell phone they charge when they have access to a generator.

Barais had been to Messi, a large coastal village less than 20 miles from the mining site, to hear a geologist talk about the project. He remembered seeing an image of a volcano on a laptop screen. "They said seabed mining cannot cause any damage, but people are not believing in what they are saying," he said. "We heard that in the world, there is no mine like this." How did he feel, I asked, about the mine being the first? "Scared," he answered. "The sea might overflow and kill us."

That concern may sound silly, but on New Ireland it—and the other mining-related fears I heard, ranging from earthquakes, tsunamis, and volcanic eruptions to massive fish kills and fundamental changes to the ocean's currents—makes a certain gloomy kind of sense. Here the precautionary principle, summed up for me by one New Irelander as "better prevent than regret," needs little explanation. The province has seen devastating landslides following logging, farmland rendered infertile by oil palm plantations, a steep decline in fish stocks after a land-based gold mine dumped toxic waste into the sea, and rising sea levels caused largely by the emissions of distant, difficult-to-imagine traffic jams and factories. New Irelanders also live in fear of the earth's tectonic power. The largest city on a neighboring island, New Britain, was nearly wiped out by a volcanic eruption in 1994, and the island was hit by another eruption this fall. How much more mysterious and forbidding than these disasters are the dark, silent depths of the sea? I parroted the line I'd heard from so many scientists: that the deep sea is like an alien world here on Earth. "Is it true?" Wilfred asked me. "There are aliens?"

When vents were first discovered, less than 40 years ago, the world hailed them as wonders. Here, in what was once thought to be a cold and featureless desert, were strange, smoking oases populated by bizarre creatures that somehow thrived without access to what was understood to be the most basic necessity of life. (Because there is no light in the deep sea, there is no photosynthesis. The energy at the base of the food web comes not from the sun but from chemical reactions.) It was an astonishing reminder of how little we understood the sea: here we were, uncovering an entirely unknown way of life on our own planet nearly a decade *after* sending astronauts to the moon.

Today the deep sea remains a world of mystery and fantasy, less mapped—and perhaps less present in our collective thoughts—than the surface of Mars. By volume, the dark regions of the ocean comprise more than 98 percent of the planet's habitat, yet we know exceptionally little about them: not the contours of their mountains and trenches, not the full life cycle of a single deep-sea species.

In Papua New Guinea opponents of seabed mining make a point of using the word *experimental* when referring to it; they also emphasize the difficulty of tracking or containing the impacts of industry in a shifting and difficult-to-study marine environment. But Nautilus and other companies argue that there are ways in which deep-ocean mining might be less damaging than terrestrial mining. Because minerals are on or fairly close to the seabed's surface, there won't be massive open-pit mines like you see on land, and therefore there will be less waste and perhaps less energy use. There won't be roads and buildings and other infrastructure left behind; everything will be mobile, ready to move on to the next site. No human communities will be displaced. And even vent ecosystems, which are naturally dynamic, won't face that much more change than they're used to. Single vents are often active for considerably less than a century due to changes in geothermal activity, becoming clogged by their own deposits or getting destroyed by a volcanic eruption. Solwara 1 is located just over a mile away from an active volcano. This close proximity means the site may disappear soon enough on its own, making deliberate decimation of it somewhat less controversial. (Hydrothermal vents are not the only place to look for minerals: mining companies are also targeting the cobalt-rich crusts of underwater mountains as well as fields of potato-sized polymetallic nodules that form in the ocean's deepest plains.)

Still, plenty of other concerns come with mining the deep ocean. Scientists worry about sediment, either kicked up off the seafloor or produced by cutting and grinding, mixing into the water and suffocating animals or disrupting filter feeders. If acidic vent fluid and metals aren't handled carefully when they're brought out of the deep, they could spill and kill reefs. Nodule mining will mean the destruction of formations that grow just a few millimeters every million years, and the mining of seamount crusts will be akin to underwater mountaintop removal on structures that serve as biological havens for fish and other animals in the open ocean.

But the biggest worry is that we may not yet know what to worry about. How do you do a risk-benefit analysis of something that's never been done before? How do you decide what's safe and what's not in a place whose workings are opaque to you? We know, for example, that the seafloor plays an important role in the way the ocean cycles heat, chemicals, and nutrients—including, crucially, carbon—but not how this process works. We're not sure how mining may compound other stressors the ocean is facing, from acidification to overfishing. The only way to know how well the deep ocean will recover from disturbance, notes Andrew Thaler, a marine ecologist who used to work in Van Dover's lab, is to disturb it.

Van Dover has publicly said that she'd prefer vent mining not to happen at all, but she is also convinced that it can't be stopped. Her best option, she believes, is to help shape how the new industry will be regulated. Given its novelty, deep-sea mining has no bad practices grandfathered in. "It's a green field," Van Dover says. "It's another frontier. We could do it right. But my sense right now is, it's a free-for-all." Some colleagues objected when she first started working with Nautilus, but Van Dover says she's recently seen other scientists working more closely with industry to develop baseline data or best practices, or to identify priority areas for protection. That shift, she says, is the result of a simple calculation with the weight of history behind it. When humans can take something we want, we usually do. And we really want minerals.

We made it to Messi the next day, but only after being turned back twice more by seemingly impassable rivers. Each time locals helped us find a way across, at one point helping us push the car out of deep mud.

Though the concerns I heard from New Irelanders were different from Van Dover's, they exposed similarly uncomfortable contradictions. In Messi a woman named Ruby William told me that the seafloor minerals are *masalai*—sacred "old things from before"—and should not be harmed. In nearly the next breath she told me that the community could get behind mining if Nautilus agreed to build a processing plant in New Ireland, creating jobs and bringing money here instead of contracting with China. But aren't the minerals sacred? I asked. She replied that Papua New Guinean workers would know which ones are sacred and which are not. (She also told me that development and money are the answer to the region's recent problems with violence and drunkenness; a few houses down, another woman told me that development and money are the source of those problems.)

It eventually became clear that one of my basic questions—if you had a choice, would you want mining?—was confusingly hypothetical. Like Van Dover, William and her family didn't see much choice. In Papua New Guinea, where most land is communally owned by extended families, villagers have real power as landowners but also know that it's not enough to stop government officials and foreigners from eventually getting what they want. And so, in the face of inevitability, they negotiate, sorting out who will get how much in royalties or necessary services.

One of the justifications companies like Nautilus offers for seabed mining is that minerals are becoming increasingly more difficult to come by on land. But people began dreaming of mining the deep ocean almost as soon as minerals were discovered there. (During the Cold War, Howard Hughes claimed to be collecting seabed minerals while he teamed with the CIA to search for a sunken Russian nuclear sub; the misdirection was enough to kick off a wave of real commercial interest.) And it's rarely mentioned that seabed mining will happen in addition to land-based mining, not instead of it, or that we could do a much better job of recycling metals and designing products to be less wasteful.

Deep-sea mining is just one version of a fairly ordinary decision: to weigh known benefits against unknown risks and choose to move ahead. Yet we squirm more than usual to learn that even the bottom of the ocean is no longer beyond the limits of human industry. This is the contradiction of the deep sea. However much it may seem to be a separate, alien world—however much we may like to think of it that way—it isn't.

Years ago Van Dover began inviting artists on dives in the hope that one of them would be able to translate the strangeness of the abyssal wilderness to those of us who will never directly witness it. She got the idea from an oceanographer named John Delaney, an early mentor. In 1991, when Van Dover was still working as an *Alvin* pilot, Delaney invited Michael Collier, a rare nonscientist, aboard a deep-sea research cruise off the Washington coast. Delaney believed the deep sea needed to be seen, and felt, by someone with Collier's area of expertise. Collier is a poet.

On the day of his scheduled dive, rough weather forced the crew to cancel *Alvin*'s descent. There was a good chance of another opportunity later in the cruise, but Collier felt he had to go home. The semester had begun, and he had classes to teach. When he told Van Dover, he remembers, "She looked at me, and she said, 'Are you *crazy?* You're going to go back? You can't go back. A week doesn't matter; another week out of your life doesn't matter. You have to stay and go see the bottom of the ocean.'"

So he stayed and made the dive, dropping through the vast darkness and the rafts of bioluminescence. When the pilot turned on the light, Collier recalls, he looked out his porthole at a field of hydrothermal vents that scientists had dubbed Krypto, Dante, and Hulk and thought to himself, This is what the beginning of the world looks like. He spotted giant clams, strange shrimp, and huge colonies of tubeworms, some beautiful and undulating, some scorched by the intensely hot vent water, looking "like wiring in your car that had melted." He felt as though he were under a spell.

It's been 23 years since that day on the seafloor, but Collier still feels the wonder of it. He's followed the advent of mining with the luxury of less ambivalence than Van Dover. "I think it's an awful idea," he tells me.

For a while Collier visited schools with a slideshow about the deep ocean, but it was years before he published a poem about his dive. The deep sea, he discovered, is as confounding for a poet as it is for a scientist —it is so bizarre, so other, so alien. "I felt this inadequacy, this essential inadequacy, about how to describe what I was seeing," he says. "As if what you had to do was create the language for it."

SAM KEAN

Phineas Gage, Neuroscience's Most Famous Patient

FROM *Slate*

1. From a Virtuous Foreman to a Sociopathic Drifter

On September 13, 1848, at around 4:30 p.m., the time of day when the mind might start wandering, a railroad foreman named Phineas Gage filled a drill hole with gunpowder and turned his head to check on his men. It was the last normal moment of his life.

Other victims in the annals of medicine are almost always referred to by initials or pseudonyms. Not Gage: his is the most famous name in neuroscience. How ironic, then, that we know so little else about the man—and that much of what we think we know, especially about his life unraveling after his accident, is probably bunk.

The Rutland and Burlington Railroad had hired Gage's crew that fall to clear away some tough black rock near Cavendish, Vermont, and it considered Gage the best foreman around. Among other tasks, a foreman sprinkled gunpowder into blasting holes and then tamped the powder down, gently, with an iron rod. This completed, an assistant poured in sand or clay, which got tamped down hard to confine the bang to a tiny space. Gage had specially commissioned his tamping iron from a blacksmith. Sleek like a javelin, it weighed 13¼ pounds and stretched 3 feet 7 inches long. (Gage stood five foot six.) At its widest the rod had a diameter of

1¼ inches, although the last foot—the part Gage held near his head when tamping—tapered to a point.

Gage's crew members were loading some busted rock onto a cart, and they apparently distracted him. Accounts differ about what happened after Gage turned his head. One says Gage tried to tamp the gunpowder down with his head still turned and scraped his iron against the side of the hole, creating a spark. Another says Gage's assistant (perhaps also distracted) failed to pour the sand in, and when Gage turned back, he smashed the rod down hard, thinking he was packing inert material. Regardless, a spark shot out somewhere in the dark cavity, igniting the gunpowder, and the tamping iron rocketed upward.

The iron entered Gage's head point first, striking below the left cheekbone. It destroyed an upper molar, passed behind his left eye, and tore into the underbelly of his brain's left frontal lobe. It then plowed through the top of his skull, exiting near the midline, just behind where his hairline started. After parabola-ing upward —one report claimed it whistled as it flew—the rod landed 25 yards away and stuck upright in the dirt, mumblety-peg-style. Witnesses described it as streaked with red and greasy to the touch, from fatty brain tissue.

The rod's momentum threw Gage backward, and he landed hard. Amazingly, he claimed he never lost consciousness. He merely twitched a few times on the ground, and was talking and walking again within minutes. He felt steady enough to climb into an oxcart, and after someone grabbed the reins and giddyapped, he sat upright for the entire mile-long trip into Cavendish. At the hotel where he was lodging, he settled into a chair on the porch and chatted with passersby. The first doctor to arrive could see, even from his carriage, a volcano of upturned bone jutting out of Gage's scalp. Gage greeted the doctor by angling his head and deadpanning, "Here's business enough for you." He had no idea how prophetic those words would be. The messy business of Gage continues to this day, 166 years later.

Most of us first encountered Gage in a neuroscience or psychology course, and the lesson of his story was both straightforward and stark: the frontal lobes house our highest faculties; they're the essence of our humanity, the physical incarnation of our highest cognitive powers. So when Gage's frontal lobes got pulped, he transformed from a clean-cut, virtuous foreman into a dirty, scary, sociopathic drifter. Simple as that. This story has had a huge influence on the scientific and popular understanding of the brain. Most uncomfortably, it implies that whenever people suffer grave damage to the frontal lobes—as soldiers might, or victims of strokes or Alzheimer's disease—something essentially human can vanish.

Recent historical work, however, suggests that much of the canonical Gage story is hogwash, a mélange of scientific prejudice, artistic license, and outright fabrication. In truth each generation seems to remake Gage in its own image, and we know very few hard facts about his post-accident life and behavior. Some scientists now even argue that, far from turning toward the dark side, Gage recovered after his accident and resumed something like a normal life—a possibility that, if true, could transform our understanding of the brain's ability to heal itself.

2. *Gage "Was No Longer Gage"*

The first story that appeared about Gage contained a mistake. The day after his accident, a local newspaper misstated the diameter of the rod. A small error, but an omen of much worse to come.

Psychologist and historian Malcolm Macmillan, currently at the University of Melbourne, has been chronicling mistakes about Gage for 40 years. He has had a peripatetic career: among other topics, he has studied disabled children, Scientology, hypnosis, and fascism. In the 1970s he got interested in Gage and decided to track down original material about the case. He turned up alarmingly little, and realized just how rickety the evidence was for most of the science about Gage.

Macmillan has been sifting fact from fiction ever since, and he eventually published a scholarly book about Gage's story and its afterlife, *An Odd Kind of Fame.* Although slowed by a faulty hip replacement—he has trouble reaching books on the bottom shelves at libraries now—Macmillan continues to fight for Gage's reputation, and he has gotten so involved with his subject that he now refers to him, familiarly, as Phineas. Above all, Macmillan stresses the mismatch between what we actually know about Gage and the popular understanding of him: "Despite there being no more than a couple hundred words attesting to how he changed, he came to dominate thinking about the function of the frontal lobes."

The most important firsthand information comes from John Harlow, a self-described "obscure country physician" who was the second doctor to reach Gage the day of the accident, arriving around 6 p.m. Harlow watched Gage lumber upstairs to his hotel room and lie down on the bed—which pretty much ruined the linens, since Gage's body was one big bloody mess. As for what happened next, readers with queasy stomachs should probably skip to the next paragraph. Harlow shaved Gage's scalp and peeled off the dried blood and brains. He then extracted skull fragments from the wound by sticking his fingers in from both ends, Chinese-finger-trap-style. Throughout this all, Gage was retching every 20 minutes, because blood and greasy bits of brain reportedly kept slipping down the back of his throat and gagging him. Incredibly, Gage never got ruffled, remaining conscious and rational throughout. He even claimed he'd be back blasting rocks in two days.

The bleeding stopped around 11 p.m., and Gage rested that night. The next morning his head was heavily bandaged and his left eyeball was still protruding a good half-inch, but Harlow allowed him visitors, and Gage recognized his mother and uncle, a good sign. Within a few days, however, his health deteriorated. His face puffed up, his brain swelled, and he started raving, at one point demanding that someone find his pants so he could go outside. His brain developed a fungal infection and he lapsed into a coma. A local cabinetmaker measured him for a coffin.

Fourteen days into the crisis, Harlow performed emergency surgery, puncturing the tissue inside Gage's nose to drain the wound. Things were touch-and-go for weeks, and Gage did lose sight in his left eye, which remained sewn shut the rest of his life. But he eventually stabilized, and in late November he returned home to Lebanon, New Hampshire—along with his tamping iron, which he started carrying around with him everywhere. In his case report, Harlow modestly downplayed his role in the recovery: "I dressed him," he wrote, "God healed him."

During his convalescence stories about Gage started circulating in newspapers, with varying degrees of accuracy. Most gave Gage the tabloid treatment, emphasizing the sheer improbability of his survival. Doctors gabbed about the case too, albeit with a dose of skepticism. One physician dismissed Gage as "a Yankee invention," and Harlow said that others, like Saint Thomas with Jesus, "refused to believe that the man had risen until they had thrust their fingers into the hole of his head."

Dr. Henry Bigelow brought Gage to Harvard Medical School for a formal evaluation in 1849. Although Bigelow treated Gage like a curiosity—he once presented Gage at a meeting along with a stalagmite "remarkable for its singular resemblance to a petrified penis"—the visit resulted in the only other detailed, firsthand account of Gage and his accident besides Harlow's. Surprisingly, Bigelow's report pronounced Gage "quite recovered in his faculties of body and mind." However, as was common in neurological exams then, Bigelow probably only tested Gage for sensory and motor deficits. And because Gage could still walk, talk, see, and hear, Bigelow concluded that his brain must be fine.

Bigelow's assessment meshed well with the medical consensus at the time, which held that the frontal lobes didn't do much—in part because people could suffer grave injuries to them and walk away. Scientists now know that parts of the frontal lobes contribute to nearly every activity inside the brain. The forefront of the lobes, called the prefrontal area, plays an especially important role in impulse control and planning.

But even today scientists have only a vague idea of how the prefrontal lobes exercise that control. And victims of prefrontal injuries can still pass most neurological exams with flying colors. Pretty much anything you can measure in the lab—memory, language, motor skills, reasoning, intelligence—seems intact in these people. It's only outside the lab that problems emerge. In particular personalities might change, and people with prefrontal damage often betray a lack of ambition, foresight, empathy, and other ineffable traits. These aren't the kind of deficits a stranger would notice in a short conversation. But family and friends are acutely aware that something is off.

Frustratingly, Harlow limited his discussion of Gage's mental status to a few hundred words, but he does make it clear that Gage changed—somehow. Although resolute before the accident, Harlow says Gage was now capricious, and no sooner made a plan than dropped it for another scheme. Although deferential to people's wishes before, Gage now chafed at any restraint on his desires. Although a "smart, shrewd businessman" before, Gage now lacked money sense. And although courteous and reverent before, Gage was now "fitful [and] irreverent, indulging at times in the grossest profanity." Harlow summed up Gage's personality changes by saying, "The equilibrium . . . between his intellectual faculties and his animal propensities seems to have been destroyed." More pithily, friends said that Gage "was no longer Gage."

As a result of this change, the railroad refused to reinstate Gage as foreman. He began traveling around New England instead, displaying himself and his tamping iron for money. This included a stint in P. T. Barnum's museum in New York—*not* Barnum's traveling circus, as some sources claim. For an extra dime skeptical viewers could "part Gage's hair and see his brain . . . pulsating" beneath his scalp. Gage finally found steady work driving a horse coach in New Hampshire.

Beyond that sketch of his activities, there's no record of what Gage did in the months after the accident—and we know even less about what his conduct was like. Harlow's case report fails to include any sort of timeline explaining when Gage's psychological symptoms emerged and whether any of them got better or worse over time. Even the specific details of Gage's behavior seem, on a closer reading, ambiguous, even cryptic. For instance, Harlow mentions Gage's sudden "animal propensities" and, later, "animal passions." Sounds impressive, but what does that mean? An excessive appetite, strong sexual urges, howling at the moon? Harlow says that Gage cursed "at times," but how often is that? And was this a saucy "hell" or "damn" here and there or something more dastardly? Harlow notes that Gage started telling his nieces and nephews wild stories about his supposed adventures. Was he confabulating here, a symptom of frontal lobe damage, or simply indulging a love of tall tales? Even the conclusion that Gage "was no longer Gage" could mean almost anything.

Indeed, it has come to mean almost anything. One reason it's hard to diagnose frontal lobe damage is that people vary quite a bit in their baseline behavior: some of us are rude, crude, cruel, flighty, or whatever naturally. To judge whether a person changed after an accident, you have to have known him beforehand. Unfortunately, no one who knew Gage intimately left any sort of statement. And with so few hard facts to constrain people's imaginations in later years, rumors began to swirl about Gage's life, until a wholly new Phineas emerged.

Macmillan summarizes this caricature of Gage as "an unstable, impatient, foul-mouthed, work-shy drunken wastrel, who drifted around circuses and fairgrounds, unable to look after himself and dying penniless." Sometimes his new traits contradicted one another: some sources describe Gage as sexually apathetic, others as promiscuous; some as hot-tempered, others as emotionally void, as if lobotomized. And some anecdotes seem like outright fabrications. In one, Gage sold the exclusive, posthumous rights to his skeleton to a certain medical school—then sold the same rights to another school, and another, skipping town and pocketing the cash each time. In another tale, a real howler, Gage lived for 20 years with the iron rod still impaled in his skull.

More uncomfortably, some scientists have questioned Gage's humanity. *Descartes' Error,* a popular book from 1994, trotted out many familiar tropes: that women couldn't stand to be in Gage's presence, that he started "drinking and brawling in questionable places," that he was a braggart and a liar and a sociopath. The neuroscientist author then got metaphysical. He speculated that Gage's free will had been compromised, and raised the possibility that "his soul was diminished, or that he had lost his soul."

People butcher history all the time, of course, for various reasons. But something distinct seems to have happened with Gage. Macmillan calls it "scientific license." "When you look at the stories told about Phineas," he says, "you get the impression that [scientists] are indulging in something like poetic license—to make the story more vivid, to make it fit in with their preconceptions." Science historian Douglas Allchin has noted the power of preconceptions as well: "While the stories [in science] are all about history—events that happened," Allchin writes, "they sometimes drift into stories of what 'should' have happened."

With Gage, what scientists think "should" have happened is colored by their knowledge of modern patients. Prefrontal lobe damage is associated with a subsequent slightly higher rate of criminal and antisocial behavior. Even if people don't sink that low, many do change in unnerving ways: they urinate in public now, blow stop signs, mock people's deformities to their faces, or abandon a baby to watch television. It's probably inevitable, Macmillan says, that such powerful anecdotes influence how scientists view Gage in retrospect: "They do see a patient and say, 'Ah, he's like what Phineas Gage was supposed to be like.'" To be clear, Harlow never reports anything criminal or blatantly unhinged about Gage's conduct. But if you're an expert on brain damage, scientific license might tempt you to read between the lines and extrapolate from "gross profanity" and "animal passions" to seedier behavior.

If repeated often enough, such stories acquire an air of truthiness. "And once you have a myth of any kind, scientific or otherwise," Macmillan says, "it's damn near impossible to get it destroyed." Macmillan especially bemoans "the degree of rigor mortis in textbooks," which reach a large, impressionable audience and repeat the same anecdotes about Gage in edition after edition. "Textbook writers are a lazy lot," he says.

Historians have also noticed, not surprisingly, that myths have more staying power when they're good stories—and Gage's is truly sensational. Once upon a time, a man with a funny name really did survive having an iron rod explode through his skull. It's tragic, macabre, bewildering—and even comes with the imprimatur of a science lesson. In contrast to other scientific fables, Gage's has an intriguing twist as well. Most other scientific myths depart from reality by inflating the heroes (usually scientists) into godlike creatures, wholly pure and wholly virtuous. Gage, meanwhile, gets demonized. He's Lucifer, fallen. Gage's myth has proved so tenacious in part because it's fascinating to watch someone break bad.

3. *The Journey of the Tamping Iron*

With the development of new scanning and computer technologies, a new chapter in Gage studies has opened in the past quarter century. Unfortunately, no one preserved Gage's brain when he died, so scientists are left examining the few remaining relics from his life instead, especially his skull and tamping iron, which are on display at the Warren Anatomical Museum at Harvard Medical School.

In six years as museum curator, Dominic Hall has become an expert on Gageanalia. He often shows the skull and tamping iron to student groups, and he finds that people don't mind hearing even graphic details about Gage's injury. "There's just something about him," Hall says.

Gage's skull and tamping iron are basically the only reason the Warren Museum still exists, says Hall, although calling it a "museum" seems generous. It's really just two rows of eight-foot-tall wooden cabinets; one sits on either side of an atrium on the fifth floor of Harvard's medical library. Surrounding the Gage artifacts are head sculptures with phrenology labels, a life mask of Samuel Taylor Coleridge, and stillborn Siamese twins, among other curios.

The left eye socket of Gage's skull, near the entry wound, looks jagged. The exit wound on top consists of two irregular holes with a patch of bone stuck between them, like a flattened wad of white gum. The tamping iron rests one shelf below the skull. Hall describes the rod as heavy, but struggles beyond that. "It's not like a baseball bat or shovel," he says, "because the weight is distributed throughout." He finally just says, "It feels *real.*" The tip of the iron looks blunted, like a slightly used crayon, and the shaft contains an inscription, in white calligraphic script, explaining Gage's case. *Phineas* is misspelled twice.

The skull's obvious entry and exit wounds have tempted several scientists to digitally recreate the journey of the tamping iron. They hope to determine what parts of the brain were destroyed, which might make Gage's deficits clearer. The sophisticated computer modeling helps scientists study normal brain function as well, but there's something undeniably splashy about recreating the most famous accident in medical history.

The best-known recreation of the accident was done by the husband-and-wife team of Antonio and Hanna Damasio, neuroscientists now at the University of Southern California. Antonio Damasio developed a famous theory of how emotions work, especially how they supplement and enhance our reasoning skills. To do so, he drew on a number of his own patients with frontal-lobe deficits. But he also drew on Gage. (Damasio, the author of *Descartes' Error,* is the scientist who described Gage as a vagrant sociopath.) The Damasios modeled Gage's accident in part to search for evidence that he suffered damage to both his left and right hemispheres, which would make any personality changes more drastic. They found what they were looking for, and the study graced the cover of *Science* in 1994.

The Damasios still stand behind their paper. But two later studies, which took advantage of higher-horsepower computers to create more accurate models of Gage's skull, have since questioned their results. In 2004 a team led by Peter Ratiu, who was then teaching neuroanatomy at Harvard and now works as an emergency doctor in Bucharest, Romania, concluded that the rod could not have crossed over the midline and damaged Gage's right hemisphere. What's more, Ratiu determined that, based on the angle of entry and lack of a broken jawbone, Gage must have had his mouth open and been speaking at the moment of impact. Ratiu's renderings of this moment—with the iron rod piercing a gaping mouth—have an unnerving quality, reminiscent of Francis Bacon's paintings of screaming popes.

In 2012 neuroimaging expert Jack Van Horn led another study on Gage's skull. In contrast to Macmillan, Van Horn refers to Phineas as "Mr. Gage." He first delved into the case while living in New Hampshire, near the old Gage farmstead on Potato Road. Van Horn now works at USC in the same department as the Damasios.

Van Horn's study sifted through millions of possible trajectories for the iron rod, he says, and ruled out all but a few "that didn't break his jaw, didn't blow his head off, and didn't do a bunch of other things." (For comparison, the Damasio study scrutinized a half-dozen trajectories.) Overall, Van Horn's work supported Ratiu's: the rod, he argues, never crossed over to the right hemisphere.

Van Horn did introduce a new wrinkle, however. He studies brain connectivity, the emerging awareness that while neurons are important to brain function, the *connections* between neurons are equally vital. Specifically, the patches of neurons that compute things in the brain (gray matter) reach their full potential only when networked together, via axon cables (white matter), to other centers of neural computation. And while Gage suffered damage to 4 percent of his gray matter, Van Horn concluded, 11 percent of his white matter suffered damage, including cables that led into both hemispheres. Overall, the injury "was much more profound than even we thought," he says.

How that damage affected Mr. Gage's behavior, though, is tough to predict. Van Horn has read Macmillan's work closely, and he says it scared him away from undue speculation. "I didn't want to piss [Macmillan] off," he jokes. Van Horn nevertheless did compare the destruction of Gage's white matter to the damage wrought by neurodegenerative diseases like Alzheimer's. Gage might even have displayed classic symptoms of Alzheimer's, he argues, such as moodiness and an inability to complete tasks. John Harlow's original case report did state that Gage's changes were "nothing like dementia," Van Horn acknowledges. But Harlow examined Gage shortly after his accident, Van Horn says, not months or years later, when such symptoms might have emerged.

Despite different interpretations, Damasio, Ratiu, and Van Horn all agree about one thing: their models are basically sophisticated guesswork. Clearly the tamping iron destroyed some brain tissue. But the flying bone shrapnel and the fungal infection would have destroyed still more tissue—and that destruction is impossible to quantify. Perhaps even more important, both the position of the brain within the skull and the location of various structures within the brain itself actually vary a lot from person to person—brains differ as much as faces do. When cataloguing brain destruction, then, millimeters matter. And no one knows which exact millimeters of tissue got destroyed in Gage.

That ignorance hasn't slowed down the speculation. Phineas Gage is reborn every generation, but as a different man: each generation reinterprets his symptoms and deficits anew. In the mid-1800s, for example, phrenologists explained Gage's profanity by noting that his "organ of veneration" had been blown to bits. Nowadays scientists cite Gage in support of theories about multiple intelligences, emotional intelligence, the social nature of the self, brain plasticity, brain connectivity—every modern neuro-obsession. Even Macmillan, after studying the end of Gage's life, has edged beyond merely debunking other people's stories and started presenting his own theory about Phineas Gage's redemption.

4. "I Knew There Was a Contradiction There"

Incredibly, after working 18 months in the horse stable in New Hampshire, Gage struck out for South America in 1852. He was seasick the whole voyage. He'd been recruited by an entrepreneur hoping to take advantage of a gold rush in Chile, and once ashore Gage resumed driving coaches, this time along the rugged, mountainous trails between Valparaiso and Santiago. You wonder how many passengers would have climbed aboard had they known about their one-eyed driver's little accident, but he did the job for seven years.

Poor health forced Gage to quit Chile, and in 1859 he caught a steamer to San Francisco, near where his family had moved. After a few months of rest, he found work as a farm laborer and seemed to be doing better, until a punishing day of plowing in early 1860 wiped him out. He had a seizure the next night over dinner. More followed, and after one particularly intense fit he died, on May 21, age 36, having survived his accident by almost a dozen years. His family buried him two days later, possibly with his beloved tamping iron.

Gage's story might have ended there—an obscure small-town tragedy, little more—if not for Dr. Harlow. He had lost track of Gage years before, but he learned the address of Gage's family in 1866 (through some unspecified "good fortune") and wrote to California for news. After milking the family for details, Harlow prevailed upon Gage's sister, Phebe, to open the grave and salvage Gage's skull in 1867. The exhumation sounded like quite a to-do, with Phebe, her husband, their family doctor, the city mortician, and even San Francisco's mayor, one Dr. Coon, all present to peek inside the coffin. Gage's family then hand-delivered the skull and tamping iron to Harlow in New York a few months later. At this point Harlow finally wrote up a full case report, which included virtually everything we know about Gage's mental status and sojourn to South America.

Most accounts of Gage's life omit all mention of Chile. Even Macmillan didn't know what to make of it for decades. But in the past few years he has become convinced that Chile holds the key to understanding Gage.

The epiphany came while, of all things, watching Queen Elizabeth's husband, Prince Philip, race coaches on television one night. Philip, an old-fashioned sportsman, drives horse coaches similar to the ones Gage did, and the intricacy of the rein-work and difficulty of the maneuvering struck Macmillan as significant. The driver controls each of his horses' reins with a different finger, for example, so even rounding a bend takes incredible dexterity. (Imagine driving a car while steering each wheel independently.) Moreover, the trails Gage drove were crowded, forcing him to make quick stops and dodges, and because he probably drove at night sometimes, he would have had to memorize their twists and drop-offs, plus watch for bandits. He also presumably cared for the horses and collected fares. Not to mention that he likely picked up a soupçon of Español in Chile. "To have someone with impulsive behavior, uncontrolled behavior, carrying out the highly skilled task of stagecoach driving," Macmillan says, "I knew there was a contradiction there."

He pursued his hunch, and after parsing and reparsing the vague chronology in Harlow's case report, Macmillan now believes that Gage's behavioral troubles were temporary and that Gage eventually recovered some of his lost mental functions. Independent evidence also supports this idea. In 2010 a computer scientist and intellectual property consultant who sometimes collaborates with Macmillan, Matthew Lena, turned up a statement from a 19th-century doctor who lived in "Chili" and knew Gage well: "He was in the enjoyment of good health," the doctor reported, "with no impairment whatever of his mental faculties." To be sure, Macmillan does not believe that Gage magically recovered everything and "became Gage" again. But maybe Gage resumed something like a normal life.

Modern neuroscientific knowledge makes the idea of Gage's recovery all the more plausible. Neuroscientists once believed that brain lesions caused permanent deficits: once lost, a faculty never returned. More and more, though, they recognize that the adult brain can relearn lost skills. This ability to change, called brain plasticity, remains somewhat mysterious, and it happens achingly slowly. But the bottom line is that the brain can recover lost functions in certain circumstances.

In particular Macmillan suggests that Gage's highly regimented life in Chile aided his recovery. People with frontal-lobe damage often have trouble completing tasks, especially open-ended tasks, because they get distracted easily and have trouble planning. But in Chile Gage never had to plan his day: prepping the coach involved the same steps every morning, and once he hit the road, he simply had to keep driving forward until it was time to turn around. This routine would have introduced structure into his life and kept him focused.

A similar regime could, in theory, help other victims of Gage-like brain damage. One gruesome paper from 1999 ("Transcranial Brain Injuries Caused by Metal Rods or Pipes over the Past 150 Years") chronicles a dozen such cases, including a drunken game of "William Tell." Another case occurred on a construction site in Brazil in 2012, when a metal bar fell five stories, pierced the back of a man's hardhat, and exited between his eyes. More commonly, people suffer brain damage on the battlefield or in car accidents. And according to a traditional reading of Gage, their prognosis was bleak. But according to Macmillan's reading, maybe not. Because if even Phineas Gage bounced back—that's a powerful message of hope.

5. *Proud, Well Dressed, Disarmingly Handsome*

Phineas Gage has probably never been more popular. Several musicians have written tributes. Someone started a blog called The Phineas Gage Fan Club, and another fan crocheted Mr. Gage's skull. YouTube contains thousands of Gage videos, including several reenactments of the accident. (One involves Barbie dolls, another Legos. Beneath one, somebody commented, inevitably, "mind=blown.") What's more, his skull has become the modern equivalent of a medieval saint's relic: the logbook at the Harvard museum has recorded pilgrims from Syria, India, Brazil, Korea, Chile, Turkey, and Australia within the past year. Comments in the book include, "An odd treat," and "Phineas Gage was on my bucket list."

More importantly, new material about Gage continues to emerge. In 2008 the first known image of Gage turned up, a sepia daguerreotype of him holding his tamping iron. (A second photo has since appeared.) The picture's owners, the collectors Jack and Beverly Wilgus, originally labeled it "the whaler," speculating that, somewhat like Ahab, the young man in it had lost his left eye to "an angry whale." But after they posted the picture on Flickr, whaling enthusiasts protested that the smooth tamping iron looked nothing like a harpoon. One commenter finally suggested it might be Gage. To check this possibility, the Wilguses compared their image to a life mask of Gage made in 1849 and found that the features lined up perfectly, including a scar on Gage's forehead. Although just one picture, it exploded the common image of Gage as a dirty, disheveled misfit. This Phineas was proud, well dressed, and disarmingly handsome.

Scientifically, Gage's legacy remains more ambiguous. His story certainly captures people's imaginations and kindles their interest in neuroscience. (Whenever I'm in mixed company and mention that I've written a book about the most fascinating injuries in neuroscience history, someone always blurts out, "Oh, like Phineas Gage!") But his story also misleads people, at least in its traditional form. Based on interviews and citations, Macmillan's revised history does seem to be gaining traction. But it's an uphill climb. "It has occurred to me [to ask] from time to time," Macmillan sighs, "what the hell I am doing working on this?"

As for the latest research on Gage—especially the brain connectivity and brain plasticity work—it seems sound. But that's really for posterity to judge. Perhaps each new theory about Gage is indeed inching us closer to the truth. On the other hand, perhaps Gage is doomed to remain a historical Rorschach blot, revealing little but the passions and obsessions of each passing era.

Because of all the uncertainty, Ratiu, the Bucharest doctor, recommends that neuroscientists stop teaching Gage. "Leave this damn guy alone," he says. (Like Gage himself, people seem to indulge in "gross profanity" when discussing his case.) But this seems unlikely. Whenever teachers need an anecdote about the frontal lobes, "you just take this ace out of your sleeve," Ratiu says. "It's just like whenever you talk about the French Revolution you talk about the guillotine, because it's so cool."

If nothing else, Macmillan says, "Phineas's story is worth remembering because it illustrates how easily a small stock of facts can be transformed into popular and scientific myth." Indeed, the myth-making continues today. "Several people have approached me with a view to develop film scripts or plays," he says. One involved Gage falling in love with a Chilean prostitute who rescues him from a life of dissolution. Another involved Gage returning to the United States, befriending and freeing a slave, then banding together with Abraham Lincoln to win the Civil War.

Another, deeper reason Gage will probably always be with us is that despite all that remains murky and obscure, his life did hint at something important: the brain and mind are one. As one neuroscientist writes, "Beneath the tall tales and fish stories, a basic truth embedded in Gage's story has played a tremendous role in shaping modern neuroscience: that the brain is the physical manifestation of the personality and sense of self." That's a profound idea, and it was Phineas Gage who pointed us toward that truth.

JOURDAN IMANI KEITH

At Risk

FROM *Orion*

THE TORRENTIAL RAIN in the first week of September pummels the youth crew's tents at night, depositing mud and sediment in the creek where they pump water for drinking. For 17 days the teenagers I recruited to build trails for the North Cascades National Park are camping during one of the heaviest storms in 100 years. The river coughs thick brown mudslides onto State Route 20, blocking road access from the west. Instead of a three-and-a-half-hour drive to pick them up, I begin a seven-hour journey eastward from Seattle through rock formations that dart out like deer from the light-green sagebrush.

Arriving in the dark, I find the Pearrygin Lake campground outside of the Old West town of Winthrop barely occupied. The warm air is without insects, so I tuck myself in on the grass behind the white 12-passenger van I have rented to pick up the crew. Lying out under the stars without the nylon canopy of my tent, I nestle into the reflection of the half-moon and backlit mountains in the lake.

Despite my comfort, I am acutely aware that I am at risk: Black. Woman. Alone. Camping. Even in the disguising cloak of moonlight shadows, I need protection. I sleep with the van keys in my pocket and practice grabbing them to push the panic button in case of danger.

The spring chinook populations in the watersheds around Route 20 are labeled "at-risk populations" when the Forest Service discusses road analysis in the Methow River subbasin and its watersheds. They are protected under the Endangered Species Act.

Protecting an endangered species means changing the practices in an entire ecosystem to safeguard their survival. It means managing the loss of their habitat, the turbidity of their waters, the surface water runoff from the streets that threatens them, and the effluents from the wastewater that disrupt their endocrine systems and, if unchecked, will cause their extinction.

Every year salmon return to the rivers where they were born. And every year I return to the birthplace of the wilderness program I developed to nurture the next generation of outdoor leaders in the cathedral peaks and azure lakes of the North Cascades. This year the wettest August followed by the heavy rains of September threatened both my crew of nascent campers and the eggs of the spawning salmon.

My youth crews are Black. Latino. Urban. This is what the woman at a party hears when I describe them. "Oh, you work with at-risk youth," she says. She doesn't hear Volunteer. College Student. Intern. Outdoor Leader. The "at-risk" label is different for youth than it is for salmon. "At-risk" isn't a protection but a limitation, a judgment, an assumption. Even when the threats to their survival are the same as to an endangered species—an unstable habitat, lack of nutrition, and a damaged social and natural ecosystem—the label leaves them at a deficit, offers no promise for protection.

In the case of the salmon, being protected as an endangered species alerts us to the fullness of their connection to a magnificent web. Their relationship to threatened indigenous cultures and to other endangered species like the majestic orca whales is valued. Their label protects them.

I tell the woman at the party, "All youth are at risk—the risks are just different." And some are endangered.

JOURDAN IMANI KEITH

Desegregating Wilderness

FROM *Orion*

I SAW SEVERAL bison the other morning as I followed the usual left and right turns that take me to the rural outskirts of Enumclaw, a town 45 minutes from my house in Seattle. The bison were actually boulders or bushes or clumps of hay that had wandered away from the stack. But I saw them as clearly as I did on mornings when I walked to my job along the lake in Yellowstone, following the gravel paths that passed dangerously close to the resting mammoths. Always, now, since leaving Yellowstone, I see shadows of wilderness wherever I go.

It is the same for my friend on the other side of the continent who, decades later, still sees the Beartooth Mountains in the jagged clouds above Philadelphia's streets. It is also true for some of the teenagers, like Michael, who I camped and worked with in the mountains and lakes of the North Cascades National Park. One early autumn I ran into Michael in our neighborhood on a south Seattle corner, which was plump with fast-food chains and drugstores. He pointed into the greasy air above a discount food outlet. "There," he said, gesturing more emphatically, "see it? It's an eagle's nest." I strained, wanting to see the shadow of the wilderness cast so close to the intersection of honking cars, but I couldn't see what he saw. Four years later he still talks about it, still insists that it was there, and now he mourns, "They have torn up the path and removed the pole where it nested."

In 1964 the Civil Rights Act and the Wilderness Act both became laws that govern the land and the people of the U.S. The Civil Rights Act required the desegregation of our public accommodations, including the "separate but equal" facilities, camping areas, and outdoor eating areas assigned to "colored" people in our national parks. The Wilderness Act protected large areas that would have been lost without laws in place to stop people from dominating every landscape. Fifty years later the cool shadow cast by these two monolithic acts makes it possible for me to occasionally enjoy outdoor experiences in remote places that would be out of reach without the combination of their protections.

Yet access is more than the permission to be somewhere formerly off limits. The people accessing recreation in the wilderness are still predominantly white, and de facto segregation exists instead of a legal one. The protections of the two acts fall short of addressing the underlying land-use practices and attitudes that have resulted in a segregated wilderness, one in which the wild is hardest to reach for the people who, for historical reasons, still have fewer of the financial assets required to get there.

While the lack of access to wild places is beginning to be recognized as an issue of inequity, the absence of more than just a shadow of wilderness in and around urban places is not.

When President Johnson signed the Wilderness Act, it legalized the segregation of wild places from the places where people remain. In doing so, it entrenched the cultural belief shared by Aldo Leopold and others that wilderness must be "segregated and preserved" from the areas where people live, including areas where indigenous people lived for thousands of years. Consequently the protection of remote wilderness areas has meant the sacrifice and disappearance of nearby wild places, which, because of their smaller size and proximity to people, are not defined or protected as wilderness.

Segregating wilderness from people creates permission to deforest and devalue the landscape where people are allowed to "remain" while falsely defining the remote landscape as "pristine." Desegregating the wilderness requires not only the laws that forbid discrimination but also the reintegration of nearby wilderness where people live.

Now largely white organizations and agencies are grappling with the dilemma of a segregated wilderness by working feverishly to get urban people out to remote places—because people will not protect what they have not enjoyed. But what if wilderness zigzagged through areas where urban people live? Then accessing the wilderness in our daily lives could be more tangible than wild shadows cast by memory.

At the signing of the Civil Rights Act, President Johnson said, "Freedom would be secure only if each generation fought to renew and enlarge its meaning." I think the same is true for wilderness.

ELI KINTISCH

Into the Maelstrom

FROM *Science*

WHEN 40 CLIMATE experts huddled in a small conference room near Washington, D.C., last September, all eyes were on an atmospheric scientist named Jennifer Francis. Three years ago Francis proposed that the warming Arctic is changing weather patterns in temperate latitudes by altering the behavior of the northern polar jet stream, the high, fast-moving river of air that snakes around the top of the world. The idea neatly linked climate change to weather, and it has resonated with the press, the public, and powerful policymakers. But that day Francis knew that many of her colleagues, including some in that room, were deeply skeptical of the idea and irritated by its high profile.

Sometimes Francis is anxious before high-pressure talks and wakes before dawn. Not this time, even though the National Academy of Sciences had assembled the group essentially to scrutinize her hypothesis. "I wasn't nervous," she recently recalled. "I was prepared for the pushback."

It came fast and hard. Just one slide into her talk, before she could show a single data point, a colleague named Martin Hoerling raised a challenge. "I'll answer that with my next figure," Francis calmly responded, her bright blue eyes wide open. Two minutes later Hoerling interrupted again, calling a figure "arbitrary." Francis, unruffled, parried—only to have Hoerling jab again.

Francis presented the evidence for her hypothesis as an orderly chain of events. "I challenged every link in the chain," recalls Hoerling, an atmospheric dynamicist at the National Oceanic and Atmospheric Administration's (NOAA's) Earth System Research

Laboratory in Boulder, Colorado. Eventually the workshop's organizer had to intervene. No more questions "so the dissertation defense can go on," nervously joked David Robinson, a climatologist at Rutgers University in New Brunswick, New Jersey, where Francis also works.

Later some attendees praised Francis's performance. "The way [Hoerling] aggressively interrupted was unusual," says Arctic scientist Walt Meier of NASA's Goddard Space Flight Center in Greenbelt, Maryland. "But she handled it very well, with grace."

Hoerling's assessment? "She was unpersuasive," he says. "The hypothesis is pretty much dead in the water."

A Stiff Headwind

Francis's hypothesis has divided colleagues ever since she first proposed it in 2011, and the divisions have only deepened as Francis became a go-to climate scientist for reporters, a marquee speaker at major conferences, and an informal consultant to John Holdren, President Barack Obama's science adviser. "It's become a shooting match over her work," says atmospheric dynamicist Walter Robinson of North Carolina State University in Raleigh. "Which side are you on?"

More than scientific bragging rights are at stake. If a warming Arctic is already affecting weather in the midlatitudes, then climate change "no longer becomes something that's remote, affecting polar bears," Meier says. Instead it's a day-to-day reality affecting billions of people—and a challenge to policymakers responsible for assessing and reducing the risks.

Yet to many Francis critics, the attention she has received is premature, a product of unusual weather in the United States and Francis's cheerfully outgoing and insistent style. Hoerling, for one, says Francis is driving "a campaign . . . This single person has been able to promulgate a conjecture into an apparent explanation of everything."

"I can't help it if the media and public are interested in my research," Francis responds. But she readily admits that all the evidence is not in and concedes that the public interest has inverted the normal life cycle of a scientific controversy. "Usually a hypothesis gets tested . . . the conclusions are solid and then it becomes news," she says. But in this case, says Stephen Vavrus, a climate modeler at the University of Wisconsin, Madison, who collaborates with Francis, "Jennifer and I have been forced into the uncomfortable position of defending—or at least explaining—our position before the scientific process has run its course."

Clouds on the Horizon

Seeking out adversity is part of Francis's character. In 1980, after her junior year in college, she and her to-be husband Peter overhauled a 14-meter sailboat named *Nunaga* and sailed around the globe, logging almost 100,000 kilometers over five years. They used a sextant to navigate and drew crude weather maps on acetate, using naval data broadcast over the radio in Morse code. Many circumnavigators make the "milk run," sticking to the relatively bucolic tropics. The pair instead pushed the limits, enduring punishing winds to round Cape Horn, dipping into the "Roaring Forties" off New Zealand, and dodging ice floes some 900 kilometers from the North Pole. At one point they struggled to fix a broken rudder during a fierce winter storm in the Tasman Sea. "We regularly placed our lives in the other's hands," Peter later wrote in a self-published volume. His wife, meanwhile, "matured from a young woman to an adult."

The Arctic foray deeply affected Francis, now 56. "I just sort of fell in love with the light up there," she says. She had been studying to be a dentist, but she switched to meteorology after returning to school, focusing subsequent graduate work on Arctic forecasting. Later, as a research professor at Rutgers, she published respected analyses of the Arctic climate with a focus on sea ice, which has lost roughly 75 percent of its fall volume since 1980.

It was a second circumnavigation of the globe, beginning in 2009, that inspired what one might call the Francis hypothesis. (This time the crew included her 12-year-old son and 14-year-old daughter.) "Gazing out at the waves, you have a lot of time to think out there," she says. Francis had been studying how a changing climate was affecting the Arctic. At sea, she flipped the equation: "I started to wonder how much the Arctic was affecting the system."

Upon return she e-mailed Vavrus in January 2011 with a

"thought I have been noodling." The Arctic is warming faster than the midlatitudes, she noted, a phenomenon known as Arctic amplification. Could that amplification—2° C more warming than the rest of the globe over the past two decades—be changing the behavior of the polar jet stream, with global consequences?

Studies dating back to the 1970s had hinted at the idea, which turns some conventional wisdom about climate change on its head. Traditionally researchers have attributed the rapid Arctic warming to local drivers such as the loss of ice and snow. In other words, the Arctic is generally seen as the victim, not the perpetrator.

Francis had doubts, however, based on her observations of the northern polar jet stream. First recognized by scientists in the 1890s, this meandering torrent, which can be up to 200 kilometers across, flows west to east some 7 to 12 km above Earth's surface at speeds of up to 400 km per hour. It forms a wavy ring around the North Pole and typically marks the border between colder, low-pressure polar air masses inside the ring (called the polar vortex) and warmer, higher-pressure air to the south.

By the time Francis and Vavrus began talking, she already suspected the jet stream was changing. In 2009 she and colleagues published a paper suggesting that its west-to-east winds were weakening, or slowing, especially after Arctic summers with less sea ice. Francis blamed Arctic warming. By reducing the air pressure gradient between the Arctic and the midlatitudes, she argued, amplification might be robbing the jet stream of the engine that drives its flow.

If so, Arctic amplification could be shaping weather farther south. Researchers have come to understand that shifts in the path and speed of the jet stream exert a powerful influence over weather in the Northern Hemisphere. When the jet meanders far to the south over North America in winter, for instance, the result is cold snaps; when it meanders far to the north, temperatures can warm well above normal.

Building on that work, Francis and Vavrus began examining changes in the amplitude of jet stream meanders, or how far the crests of its bends reach north and south. Combing through atmospheric data, they found that the amplitudes in the fall and winter had increased by roughly 150 kilometers over the past 30 years, as the Arctic warmed. The northern peaks (called ridges by meteorologists) tended to stretch farther toward the Arctic, they found. The southern dips, known as troughs, were apparently affected less, but overall the jet stream seemed to be becoming more sinuous.

Like the weakening of the winds themselves, that increased "waviness," as some researchers call it, would tend to slow the eastward movement of weather patterns. The result: weather conditions of all sorts—dry periods and warm spells, or storms and cold snaps—would persist. In North America, for example, large pools of Arctic air would linger longer over the continent, as they did during this past winter.

The end result of all that slower motion, Francis believes, is more persistent weather that could be more extreme—and she said as much at a meeting of the American Geophysical Union in San Francisco in late 2011. As examples she pointed to weather events of the previous two years—long, snowy winters in the eastern United States and Europe, a lengthy Texas heat wave, and a record-breaking rainy spell in the U.S. Northeast. All were "consistent" with her analysis, Francis said.

Checking the Barometer

After that talk, "I was mobbed," Francis recalls. A few months later, in March 2012, Francis and Vavrus formally outlined their idea in *Geophysical Research Letters* (*GRL*). The timing was uncanny: temperatures in the United States were skyrocketing again. Within weeks, the *New York Times* ran a front-page story on the "surreal heat wave" and a subsequent frigid cold snap. Francis was the first scientist quoted. "The question really is not whether the loss of the sea ice can be affecting the atmospheric circulation on a large scale," she said. "The question is, how can it not be?"

"And then my life changed," Francis says. Before the *GRL* paper appeared, she estimates she spent just one quarter of her time working on outreach and communication. Soon after, that fraction rose to 80 percent. Since 2011 she has logged more than 150 media mentions and speaking engagements. She's an articulate scientist, after all, with a surprising take on a topic that everyone loves to talk about: the weather.

40-Knot Winds

As Francis has accumulated media appearances, however, opposition to the hypothesis has grown steadily among researchers. In early 2011, for instance, she and Vavrus submitted a proposal to analyze data and model the phenomenon to the National Science Foundation's (NSF's) climate dynamics division. It got generally positive reviews, although it didn't make the funding cut. A year later, however, despite revisions, "the reviews of our second attempt were much worse," Francis says. "That's when we realized there was a backlash." (NSF's Arctic science program ultimately funded the work.)

Criticism is coming from three directions. First, scientists have challenged the pair's analysis of historical data, questioning whether it really shows that the polar jet stream's west-to-east winds are slowing and its meanders stretching. Last year in *GRL*, for example, climate modeler James Screen of the University of Exeter in the United Kingdom and a colleague reported that they had measured the meanders and found few statistically significant changes. "It could easily just be natural variability," Screen says. The pair did find a reduction in the size of the jet stream's vertical waves, which rise and fall perpendicular to Earth's surface. But that is inconsistent with the Francis hypothesis, they say, because it would translate into fewer temperature extremes at any specific latitude. Last year climate dynamicist Elizabeth Barnes of Colorado State University, Fort Collins, also analyzed the data and concluded that the Francis and Vavrus findings were an "artifact of [their] methodology."

Climate modelers also have offered heckles, saying their computer simulations have mostly failed to confirm the hypothesis. In their models they've dialed up future greenhouse warming or reduced Arctic sea ice—both factors that should amp up Arctic amplification—but failed to produce a slower, more meandering jet stream. And models that simply reproduce existing conditions, Screen says, have to run for the equivalent of more than "sixty years before I start to see anything" similar to Francis's observations.

The most vociferous critiques, however, have come from researchers who study atmospheric dynamics, or the many mechanisms that jostle and shape air masses. Given the Arctic's relatively puny influence over the planet's atmospheric energy flows, the notion that it can alter the jet stream "is just plain wrong," says dynamicist Kevin Trenberth of the National Center for Atmospheric Research in Boulder. The more likely culprit, he says, is natural variability driven by the tropics, where Earth gets its largest input of solar energy.

Such variability, Trenberth says, could explain the jet stream's giant curvy shape this past January, which brought record chill to the southeastern United States and warm temperatures to Alaska and made "polar vortex" a household term. At the time, a massive amount of so-called latent heat was accumulating in the tropical Pacific, Trenberth notes, in an incipient El Niño event. Parcels of warm air from the tropics may have forced the jet stream northward in one place, causing it to meander southward farther east. "It may not be that Arctic amplification is causing a wavier jet stream, it may be that a wavier jet stream is causing Arctic amplification," he says.

"I understand that people would be skeptical," Francis says. "It's a new paradigm." But she counsels patience. She notes that evidence of Arctic amplification itself has emerged from the statistical noise only in the last 15 or so years, so it may take time for the changes to the jet stream to become statistically significant. And she believes the modeling experiments that fail to simulate a more meandering jet stream are biased, because they don't include sufficiently robust Arctic amplification.

Such arguments have persuaded some colleagues to at least wait and see. Oceanographer James Overland of NOAA's Pacific Marine Environmental Laboratory in Seattle, Washington, for example, says, "I find the tropical explanation for the recent behavior of the jet stream no less implausible than the Arctic one." And he suspects that as data accumulate, the dynamicists will come to gain a greater appreciation for the Arctic's role.

Batten the Hatches

Scientists may debate the reasons for this past January's cold spell in the eastern United States, but one clear effect was to direct more attention to the Francis hypothesis. In early January, Francis was at home in Marion, Massachusetts, responding to a blizzard of e-mails from reporters, her cat Kessie on her lap. Then came a message from an unexpected address. "I have been following with interest your work," wrote White House science adviser Holdren.

"I fell off my chair first, and started breathing again," Francis recalls.

Holdren wanted to learn more about her research, he wrote, and any relevant unpublished work. "I work for a very smart President," he explained in a follow-up message. "I don't go near him with any chart I can't completely explain!" Francis says she "fell off my chair again."

That same day, January 8, Holdren appeared in a YouTube video produced by the White House in which he essentially endorsed her hypothesis. "There will be continuing debate," he said over clips of stark Arctic ice and blowing snowdrifts. "But I believe the odds are that we can expect, as a result of global warming, to see more of this pattern of extreme cold in the midlatitudes."

"I was blown away that he was so convinced," Francis says. She was hardly alone. When dynamicist John Wallace of the University of Washington, Seattle, saw the video, he was appalled. He quickly recruited four colleagues to pen an op-ed challenging Holdren's message. "Normally I don't have time to write letters to newspapers," says Wallace, who didn't mention Holdren by name in the piece. The Francis hypothesis "deserves a fair hearing," the quintet wrote in a letter that eventually appeared in *Science.* "But to make it the centerpiece of the public discourse on global warming is inappropriate." Later, in another article, Wallace warned: "When the public becomes confused, the carefully considered scientific consensus [on climate] becomes vulnerable to attack."

"That really hurt," Francis says. But she won't back down from speaking out. The discussion "in the media has really galvanized some people to realize climate change is happening right now," she says. An oceanographer who collaborates with Francis, Charles Greene of Cornell University, agrees. "When we see something happening," he says, "we should put it out there."

Francis has taken to that mission with zeal. She sends reporters long e-mails answering their questions, carefully tracks her media hits, and continually rehones and rehearses her presentations. After a recent talk at the annual meeting of AAAS (which publishes *Science*), she was approached by Lewis Branscomb, 87, a U.S. science policy luminary with lengthy experience in Washington. "That was the best general-audience lecture I have ever heard," Branscomb told her.

She's also learned from some mistakes. In an incident this past January, Francis asserted that the "intent" of one of her critics—Colorado State's Barnes—"seems less than objective." That personal assault was included in a long technical e-mail to a weather blogger, and the comments drew a public scolding from prominent climate blogger Judith Curry of the Georgia Institute of Technology in Atlanta. Francis promptly apologized. "That e-mail was written at five o'clock in the morning while I was on a college tour with my daughter," Francis says. "Usually I like to let things simmer."

She has conceded some scientific points too. She largely dropped one part of her hypothesis—that a curvier jet stream is leading to more atmospheric "blocking"—after Barnes published an analysis challenging the idea.

Francis predicts that "within a few years, as Arctic amplification continues, we will have enough data to know whether or not we're right." In the meantime she is as comfortable as ever weathering the squalls. "I've developed a thicker skin," she says. At a recent meteorology conference, she suggested that curvier jet streams would steer more future Atlantic hurricanes west, along the path taken by Superstorm Sandy in October 2012. That contention drew fire from critics, including modelers whose work suggests the opposite. Her reaction? "That was kind of fun because people were irately skeptical," she says.

To put it all in perspective, Francis thinks back to the more serious dangers she faced at age 22 aboard the *Nunaga.* "Maybe this acceptance of higher risk was something I was more comfortable with than most, and maybe it translated to my research as having more confidence in myself—my judgment and my ability," she writes in an e-mail. The title of the book, which documents that life-changing journey, seems apt these days in more ways than one: *A Path to Extremes.*

ELIZABETH KOLBERT

The Big Kill

FROM *The New Yorker*

IN THE DAYS—perhaps weeks—it had spent in the trap, the stoat had lost most of its fur, so it looked as if it had been flayed. Its exposed skin was the deep, dull purple of a bruise, and it was coated in an oily sheen, like a sausage. Stoat traps are often baited with eggs, and this one contained an empty shell. Kevin Adshead, who had set the trap, poked at the stoat with a screwdriver. It writhed and squirmed, as if attempting to rise from the dead. Then it disgorged a column of maggots.

"Look at those teeth," Adshead said, pointing with his screwdriver at the decomposing snout.

Adshead, who is 64, lives about an hour north of Auckland. He and his wife, Gill, own a 3,500-acre farm, where for many years they raised cows and sheep. About a decade ago they decided they'd had enough of farming and left to do volunteer work in the Solomon Islands. When they returned, they began to look at the place differently. They noticed that many of the trees on the property, which should have been producing cascades of red flowers around Christmastime, instead were stripped bare. That was the work of brushtail possums. To save the trees, the Adsheads decided to eliminate the possums, a process that involved dosing them with cyanide.

One thing led to another, and soon the Adsheads were also going after rats. With them the preferred poison is an anticoagulant that causes internal hemorrhaging. Next came the stoats, or, as Americans would say, short-tailed weasels. To dispatch these, the Adsheads lined their farm with powerful traps, known as DOC

200s, which feature spring-controlled kill bars. DOC 200s are also helpful against ferrets, but the opening is too small for cats, so the Adsheads bought cat traps, which look like rural mailboxes except that inside, where the letters would go, there's a steel brace that delivers an uppercut to the jaw.

The Adsheads put out about 400 traps in all, and they check them on a regular rotation. When I visited, on a bright blue day toward the end of the Southern Hemisphere winter, they offered to show me how it was done. They packed a knapsack of supplies, including some eggs and kitty treats, and we set off.

As we tromped along, Kevin explained his trapping philosophy. Some people are fastidious about cleaning their traps of bits of rotted stoat. "But I'm not," he said. "I like the smell in there; it attracts things." Often he experiments with new techniques; recently he'd learned about a kind of possum bait made from flour, molasses, and cinnamon, and Gill had whipped up a batch, which was now in the knapsack. For cats he'd found that the best bait was Wiener schnitzel.

"I slice it thin and I tie it over the trigger," he told me. "And what happens with that is it starts to dry out and they still go for it."

I'd come to watch the Adsheads poke at decaying stoats because they are nature lovers. So are most New Zealanders. Indeed, on a per capita basis, New Zealand may be the most nature-loving nation on the planet. With a population of just four and a half million, the country has some 4,000 conservation groups. But theirs is, to borrow E. O. Wilson's term, a bloody, bloody biophilia. The sort of amateur naturalist who in Oregon or Oklahoma might track butterflies or band birds will, in Otorohanga, poison possums and crush the heads of hedgehogs. As the coordinator of one volunteer group put it to me, "We always say that for us, conservation is all about killing things."

The reasons for this are in one sense complicated—the result of a peculiar set of geological and historical accidents—and in another quite simple. In New Zealand anything with fur and beady little eyes is an invader, brought to the country by people—either Maori or European settlers. The invaders are eating their way through the native fauna, producing what is, even in an age of generalized extinction, a major crisis. So dire has the situation become that schoolchildren are regularly enlisted as little exterminators. (A recent blog post aimed at hardening hearts against cute little fuzzy things ran under the headline "Mrs. Tiggy-Winkle, Serial Killer.")

Not long ago New Zealand's most prominent scientist issued an emotional appeal to his countrymen to wipe out all mammalian predators, a project that would entail eliminating hundreds of millions, maybe billions, of marsupials, mustelids, and rodents. To pursue this goal—perhaps visionary, perhaps quixotic—a new conservation group was formed this past fall. The logo of the group, Predator Free New Zealand, shows a kiwi with a surprised expression standing on the body of a dead rat.

New Zealand can be thought of as a country or as an archipelago or as a small continent. It consists of two major islands—the North Island and the South Island, which together are often referred to as the mainland—and hundreds of minor ones. It's a long way from anywhere, and it's been that way for a very long while. The last time New Zealand was within swimming distance of another large landmass was not long after it broke free from Australia, 80 million years ago. The two countries are now separated by the 1,200-mile-wide Tasman Sea. New Zealand is separated from Antarctica by more than 1,500 miles and from South America by 5,000 miles of the Pacific.

As the author David Quammen has observed, "Isolation is the flywheel of evolution." In New Zealand the wheel has spun in both directions. The country is home to several lineages that seem impossibly outdated. Its frogs, for example, never developed eardrums but, as if in compensation, possess an extra vertebra. Unlike frogs elsewhere, which absorb the impact of a jump with their front legs, New Zealand frogs, when they hop, come down in a sort of belly flop. (As a recent scientific paper put it, this "saltational" pattern shows that "frogs evolved jumping before they perfected landing.") Another "Lost World" holdover is the tuatara, a creature that looks like a lizard but is in fact the sole survivor of an entirely separate order—the Rhynchocephalia—which thrived in the early Mesozoic. The order was thought to have vanished with the dinosaurs, and the discovery that a single species had somehow managed to persist has been described as just as surprising to scientists as the capture of a live *Tyrannosaurus rex* would have been.

At the same time New Zealand has produced some of nature's

most outlandish innovations. Except for a few species of bats, the country has no native mammals. Why this is the case is unclear, but it seems to have given other groups more room to experiment. Weta, which resemble giant crickets, are some of the largest insects in the world; they scurry around eating seeds and smaller invertebrates, playing the part that mice do almost everywhere else. *Powelliphanta* are snails that seem to think they're wrens; each year they lay a clutch of hard-shelled eggs. *Powelliphanta* too are unusually big—the largest measure more than three and a half inches across—and in contrast to most other snails, they're carnivores and hunt down earthworms, which they slurp up like spaghetti.

New Zealand's iconic kiwi is such an odd bird that it is sometimes referred to as an honorary mammal. When it was first described to English naturalists, in 1813, they thought it was a hoax. Kiwi are covered in long, shaggy feathers that look like hair, and their extended, tapered beaks have nostrils on the end. They are around the size of chickens but lay eggs that are 10 times as large, and it usually falls to the male, Horton-like, to hatch them.

New Zealand's biggest oddballs were the moa, which, in a feathers-for-fur sort of way, stood in for elephants and giraffes. The largest of them, the South Island giant moa, weighed 500 pounds, and with its neck outstretched could reach a height of 12 feet. Moa fed on New Zealand's native plants, which, since there were no mammalian browsers, developed a novel set of defenses. For instance, some New Zealand plants have thin, tough leaves when they are young, but when they mature—and grow taller than a moa could chomp on—they put out leaves that are wider and less leathery. The Australian naturalist Tim Flannery has described New Zealand's avian-dominated landscape as a "completely different experiment in evolution." It shows, he has written, "what the world might have looked like if mammals as well as dinosaurs had become extinct 65 million years ago, leaving the birds to inherit the globe." Jared Diamond once described the country's native fauna as the nearest we're likely to come to "life on another planet."

With about half the population of New Jersey, New Zealand is the sort of place where everyone seems to know everyone else. One day, without quite understanding how the connection had been made for me, I found myself in a helicopter with Nick Smith, who was then the country's minister for conservation.

Smith, who is 49, has a ruddy face and straw-colored hair. He is a member of the country's center-right National Party and calls himself a "Bluegreen." (Blue is the National Party's color.) He got interested in politics back in the 1970s, when, as an exchange student in Delaware, he met Joe Biden. We set off from Smith's district office, in the city of Nelson, on the northern tip of the South Island, and drove out to the helicopter pad in his electric car.

"When the first settlers came here, they tried to create another England," Smith told me. "We were Little Britain. The comment about us was, we were more British than the British. And as part of the maturing of New Zealand, there's the question, What do you connect your nationhood to? You know, for America it's very much the flag, the Constitution, those sorts of things. The connection with species that are unique to New Zealand is increasingly part of our national identity. It's what we are as New Zealanders, and I make no bones of the fact that the government is keen to encourage that. You need some things for a country to hold together."

He went on, "I say to people, If you want your grandkids to see kiwi only in sanctuaries, well, that's where we're headed. And that's why we need to use pretty aggressive tools to try to turn this around."

My visit happened to coincide with the application of one of these aggressive tools. The country's Department of Conservation was conducting a massive aerial drop of a toxin known as 1080. (The key ingredient in 1080, sodium fluoroacetate, interferes with energy production on a cellular level, inflicting what amounts to a heart attack.) New Zealand, which has roughly one tenth of 1 percent of the world's land, uses 80 percent of its 1080. This year's drop—the department was planning to spread 1080 over nearly two million acres—had been prompted by an unusually warm winter, which had produced an exceptionally large supply of beech seed, which in turn had produced an explosion in the number of rats and stoats. When the beech seed ran out, the huge cohort of predators was expected to turn its attention to the native fauna. Smith had approved the 1080 operation, which had been dubbed Battle for Our Birds, but the timing of it troubled him; owing to the exigencies of rat biology, the drop had to take place right around the time of a national election.

"If you ask the cynical politics of it, people don't like poisons but they like rats even less," he told me. "And so I've been doing a few quite deliberate photo opportunities with buckets of rats."

On this particular day, Smith was attending a more cheerful sort of photo op—one with live animals. When we arrived at the helicopter pad, three other people were already there, all associated with a privately funded effort to restore one of the country's most popular national parks, named for Abel Tasman. (In 1642, Tasman, a Dutch explorer, was the first European to reach New Zealand—though he didn't quite reach it, as four of his sailors were killed by Maori before they could land.) Smith and I got into the helicopter next to the pilot, the other three climbed into the back, and we took off. We flew over Tasman Bay and then over the park, which was studded with ghostly white trees.

"We like to see all those dead trees," Devon McLean, the director of the restoration project, announced cheerfully into his headset. He explained that the trees were invasive pines, known in New Zealand as "wilding conifers." I had a brief vision of scrawny seedlings rampaging through the forest. Each dead tree, McLean said, had been individually sprayed with herbicide. He was also happy to report that the park had recently been doused with 1080.

After about half an hour we landed on a small island named Adele, where we were greeted by a large sign: HAVE YOU CHECKED FOR RATS, MICE AND SEEDS? A few years ago, after an intensive campaign of poisoning and trapping, Adele was declared "pest-free." The arrival of a single pregnant rat could undo all that work; hence the hortatory signage. In another helicopter the conservation department was going to deliver two or three dozen representatives of one of New Zealand's rarest species, the South Island saddleback. The birds would be released onto Adele, where, it was hoped, in the absence of rats, they would multiply.

More people began to arrive by boat—reporters, representatives of several regional conservation groups, members of the local Maori *iwi,* or tribe. By this point it was drizzling, but there was a festive mood on the beach, as if everyone were waiting for a celebrity. "Hardly any New Zealanders have ever seen a saddleback," a woman whispered to me.

In anticipation of the birds' appearance, speeches were offered in Maori and English. "It has taken us *Pakeha* New Zealanders a little while to gain an appreciation of what is special here and to really be committed to its protection and survival," Smith said, using the Maori word for "European." "It's kind of scary to think of South Island saddlebacks—there are only about six hundred that exist on the planet."

Finally the birds arrived, in a helicopter that had been loaned for the occasion by a wealthy businessman. Three crates were unloaded onto the beach, and Smith and a pair of local dignitaries were given the honor of opening them. South Island saddlebacks are glossy black, with patches of rust-colored feathers around their middles and little orange wattles that make them look as though they're smiling. They are another example of an ancient lineage that persists in New Zealand, and they have no close relatives anywhere else in the world. The birds hopped out of their crates, flew into the bush, and were gone.

There are two quick ways to tell a Norway rat from a ship rat. One is to look at the ears. Ship rats have large ears that stick out from their heads; Norway rats' ears are shorter and less fleshy. The other is to look at their tails. Again, Norway rats' are shorter. With a ship rat, if you take its tail and fold it over its body—here it obviously helps if the rat is dead—it will extend beyond its nose.

These and many other facts about rats I learned from James Russell, an ecologist at the University of Auckland. Russell's office is filled with vials containing rat body parts in various stages of decomposition, and he also keeps a couple of dead rats at home in his freezer. Wherever he goes, Russell asks people to send him the tails of rats that they have trapped, and often they oblige. Russell then has the rats' DNA sequenced. Eventually he hopes to be able to tell how all of New Zealand's rat populations are related.

"I would be inclined to say rats are our biggest problem," Russell told me. "But I have colleagues who spend their career on stoats, and colleagues who spend their career on cats. And they open all their talks with 'Stoats are the biggest problem' or 'Cats are the biggest problem.'"

Russell, who is 35, is a slight man with tousled brown hair and a cheerful, let's-get-on-with-it manner which I eventually came to see as very New Zealand. I ended up spending a lot of time with him, because he volunteered to guide me along some of the country's windier back roads.

"New Zealand was the last large landmass on earth to be colonized," he told me one day as we zipped along through the midsection of the North Island. "And so what we saw was the tragedy of human history playing out over a short amount of time. We're only ten years behind a lot of these things; that's as compared to countries where you're hundreds or thousands of years behind the catastrophe."

New Zealand's original settlers were the Maori, Polynesians who came around the year 1300, probably from somewhere near the Society Islands. By that point people had already been living in Australia for some 50,000 years. They'd been in continental North America for at least 10,000 years, and in Hawaii, which is even more remote than New Zealand, for more than 500 years. In each case, it's now known, the arrival of humans precipitated a wave of extinctions; it's just that, as Russell points out, these "tragedies" were not recorded by the people who produced them.

When the Maori showed up, there were nine species of moa in New Zealand, and it was also home to the world's largest eagle—the Haast's eagle—which preyed on them. Within a century or two the Maori had hunted down all of the moa, and the Haast's eagle too was gone. A Maori saying, *"Ka ngaro i te ngaro a te moa,"* translates as "Lost like the moa is lost."

In their ships the Maori also brought with them Pacific rats, or *kiore*. These were New Zealand's first introduced mammals (unless you count the people who brought them). The Maori intended to eat the *kiore*, but the rats multiplied and spread far faster than they could be consumed, along the way feasting on weta, young tuatara, and the eggs of ground-nesting birds. In what in evolutionary terms amounted to no time at all, several species of New Zealand's native ducks, a couple of flightless rails, and two species of flightless geese were gone.

The arrival of British settlers, in the middle of the 19th century, brought in more—many more—new invaders. Some of them, like the Norway rat and the ship rat, were stowaways. Others were introduced deliberately, in an effort to make New Zealand feel more like home. The "importation of those animals and birds, not native to New Zealand," an 1861 act of the colonial parliament declared, would both "contribute to the pleasure and profit of the inhabitants" and help maintain "associations with the Old Country." What were known as "acclimatization societies" sprang up in every region. Among the many creatures the societies tried to "acclimatize" were red deer, fallow deer, white-tailed deer, sika deer, tahr, chamois, moose, elk, hedgehogs, wallabies, turkeys, pheasants, partridges, quail, mallards, house sparrows, blackbirds, brown trout, Atlantic salmon, herring, whitefish, and carp. Brushtail possums were specially imported from Australia, in an attempt to start a fur industry.

Not all the new arrivals took; others took all too well. By the perverse logic of such affairs, some of the most disastrous introductions were made in an effort to control previous disastrous introductions. Within a few decades of their importation, European rabbits had overrun the countryside, and in 1876 an act was passed to "provide for the Destruction of Rabbits in New Zealand." The act had no perceptible impact, so stoats and ferrets were released into the bush in the hope that they would be more effective.

"Our forebears tried most experiments that could possibly be conceived and some that would be difficult for anyone with any knowledge of ecology to seriously contemplate," Robert McDowall, a New Zealand naturalist, has written in a history of introduction efforts. The combined effect of all these "experiments," particularly the introduction of predators, like stoats, has been ongoing devastation. Roughly a quarter of New Zealand's native bird species are now extinct, and many of those which remain are just barely hanging on. If current trends continue, it is predicted that within a generation or two the land of Kiwis will be without kiwi.

"The defence of isolation for remote islands has no fallback position," John McLennan, a New Zealand ecologist, has written. "It is all or nothing, akin to virginity, with no intermediate state."

Sirocco, a sexually dysfunctional kakapo, normally lives alone on his own private island. On occasion, though, he is brought, with great fanfare, to the mainland, and this is where James Russell and I set out to meet him, at a special mountaintop reserve ringed by a 29-mile fence. The fence is 7 feet high and made of steel mesh with openings so narrow an adult can't stick a pinkie through. At the base of the fence, an 18-inch apron prevents rats from tunneling under; on top, an outwardly curved metal lip stops possums and feral cats from clambering over. To get inside, human visitors have to pass through two sets of gates, an arrangement that made me think of a maximum-security prison turned inside out.

Kakapo—the name comes from the Maori, meaning "parrot of the night"—are nocturnal, so all audiences with Sirocco take place after dark. Russell and I joined a group of about 20 other visitors, who had paid $40 apiece to get a peek at the bird. Sirocco was hopping around in a dimly lit plate-glass enclosure. He was large —about the size of an osprey—with bright green-and-brown feathers and a bulbous, vaguely comical beak. He gazed through the glass at us and gave a sharp cry.

"He's very intense, isn't he?" the woman next to me said.

Alone among parrots, kakapo are flightless, and alone among flightless birds, they're what's known as lek breeders. During mating season a male will hollow out a little amphitheater for himself, puff up his chest, and let out a "boom" that sounds like a foghorn. Kakapo, which can live to 80, breed only irregularly, in years when their favorite foods are in good supply.

Kakapo once were everywhere in New Zealand. In the late 19th century they were still plentiful in rugged areas; Charlie Douglas, an explorer who climbed some of the steepest mountains of the South Island, described them standing "in dozens round the camp, screeching and yelling like a lot of demons." But then their numbers crashed. By the 1970s there was just one small population remaining, and it was threatened by feral cats. In the 1980s every individual that could be caught was captured and "translocated." Today there are 126 kakapo left, and all of them, save Sirocco, live on three remote, predator-free islands: Little Barrier, Codfish, and Anchor.

Sirocco's chaperone, a Department of Conservation ranger named Alisha Sherriff, had brought along a little metal container, which she passed among the visitors. Inside was half a cup's worth of Sirocco's shit.

"Have a good sniff," she suggested.

"It's earthy!" one woman exclaimed.

"I think it smells smoky, with notes of honey," Sherriff said. When the container came to me, I couldn't detect any honey, but the bouquet did strike me as earthy, with hints of newly mown hay. Sherriff had also brought along a zip-lock bag with some of Sirocco's feathers. These too had a strong, sweetish scent.

New Zealand birds tend to smell, which was not a problem when the islands' top predators were avian, since birds hunt by sight. But as mammals hunt with their noses, it's become yet another liability. (A few years ago a Christchurch biologist was awarded a $400,000 grant to investigate the possibility of developing some sort of "deodorant" for ground-nesting birds.)

Sherriff explained that Sirocco had been born on Codfish Island in 1997, but as a chick he had come down with a respiratory infection and so had been removed from his mother and raised in isolation. By the time he was well enough to rejoin the other kakapo, he'd decided he preferred people. During breeding season he'd try to mate with the rangers on Codfish while they were walking to the outhouse. A special barrier was built to try to prevent the encounters, but Sirocco turned out to be too determined.

"When you've got a three-kilo bird crashing through the bush in the middle of the night, there's quite a high risk of people lashing out unintentionally," Sherriff said. It had been decided that for Sirocco's own safety, he'd have to be moved. He now lives hundreds of miles from Codfish, on an island whose name the Department of Conservation won't disclose. For the sake of genetic diversity, Sirocco's semen had been collected, by means of what Sherriff described as a "delicate massage," but his sperm count had proved too low to attempt artificial insemination.

After about half an hour with Sirocco, we were hustled out to make room for the next tour group. Russell and I passed back through the gates and drove down to a restaurant at the base of the mountain, where we'd arranged to have dinner with Matt Cook, the reserve's natural-heritage manager. Cook told us that it had taken teams of exterminators three years to eliminate mammals from inside the fence and that they'd never managed to finish off the mice. The entire perimeter was wired so that when, say, a section of fence was hit by a falling branch, a call automatically went out to a maintenance crew.

"This always happens at 3 a.m. on a Saturday morning," he said. If the fence was breached, Cook reckoned, the crew had about 90 minutes to repair it before rats would find the opening and sweep back in.

Today invasive species are everywhere. No matter where you are reading this, almost certainly you are surrounded by them. In the northeastern United States, common invasive plants include burdock, garlic mustard, purple loosestrife, and multiflora rose; invasive birds include starlings, rock pigeons, and house sparrows; and invasive insects include Japanese beetles, gypsy moths, and hemlock woolly adelgid. Texas has more than 800 nonnative plant species, California at least 1,000. Even as New Zealand has been invaded by nonnative species, its native species have invaded elsewhere. *Potamopyrgus antipodarum,* or, as it is more commonly known, the New Zealand mud snail, is a tiny aquatic snail about the size of a grain of rice. It can now be found in Europe, Asia, Australia, the Middle East, and the American West, and it has proved such a successful transplant that in some parts of the world it reaches densities of half a million snails per square yard. In an irony perhaps only Kiwis can appreciate, a recent study in Utah found that mud snails were threatening populations of rainbow trout, a fish imported at great expense to New Zealand in the 1880s and now considered an invader there.

The project of reshuffling the world's flora and fauna, which began slowly with the spread of species like the Pacific rat and sped up thanks to the efforts of acclimatization societies, has now, with global trade and travel, accelerated to the point that on any given day something like 10,000 species are being moved around just in the ballast water of supertankers. Such is the scale of this remix that biologists have compared it to reassembling the continents. Two hundred million years ago, all of the world's landmasses were squished together into a single giant supercontinent, Pangaea. We are, it's been suggested, creating a "new Pangaea."

One response is simply to accept this as the planet's destiny. Yes, the invaders will inevitably choke out some local species, and there will be losses, especially on islands, where, unfortunately, much of the world's diversity resides. But people aren't going to stop shipping goods and they aren't going to stop traveling; therefore we're just going to have to learn to live in a Pangaea of our own making. Meanwhile, who even knows at this point what's native and what's not? Many species that people think of as a natural part of the landscape are really introductions that occurred before recent memory. For instance, the ring-necked pheasant, the official state bird of South Dakota, is an import from China.

"It is impractical to try to restore ecosystems to some 'rightful' historical state," a group of American researchers wrote a few years ago, in *Nature.* "We must embrace the fact of 'novel ecosystems' and incorporate many alien species into management plans, rather than try to achieve the often impossible goal of eradicating them."

New Zealanders are nothing if not practical. They like to describe the national mindset as "the No. 8 wire mentality"; for much of the country's history, No. 8 wire was used to fix livestock fences and just about everything else. Nevertheless, Kiwis refuse to "embrace" novel ecosystems. In the past few decades they have cleared mammalian predators from 117 offshore islands. The earliest efforts involved tiny specks, like Mokopuna, or, as it is also known, Leper Island, which is about the size of Gramercy Park. But more recently they've successfully de-ratted much larger islands, like Campbell, which is the size of Nantucket. With its predator-free islands and its fenced-in reserves and its massive poison drops from the air, New Zealand has managed to bring back from the very edge of oblivion several fantastic birds, including the kakapo, the South Island saddleback, the Campbell Island teal, and the black robin. At its lowest point, the black robin was down to just five individuals, only one of which—a bird named Old Blue—was a fertile female. (When Old Blue died, her passing was announced in parliament.)

Meanwhile, by tackling larger and larger areas, New Zealanders have expanded the boundaries of what seems possible, and they increasingly find their skills in demand. When, for example, Australia decided to try to get rid of invasive rodents on Macquarie Island, roughly halfway between Tasmania and Antarctica, it hired a New Zealander to lead the effort, and when the U.S. National Park Service decided to get rid of pigs on Santa Cruz Island, off the coast of Southern California, it hired Kiwis to shoot them. The largest rat-eradication effort ever attempted is now in progress on South Georgia Island, a British territory in the South Atlantic with an area of nearly a million acres. A New Zealand helicopter pilot was brought in to fly the bait-dropping missions. One day when I was driving around with James Russell, he got an e-mail from Brazil: the government wanted to hire him to help it get rid of rats on the Fernando de Noronha archipelago, off Recife. David Bellamy, a British environmentalist and TV personality, has observed that New Zealand is the only country in the world that has succeeded in turning pest eradication into an export industry.

*

The idea of ridding all of New Zealand of its mammalian predators was proposed by Paul Callaghan, a world-renowned physicist, in a speech delivered in Wellington in February 2012. In scientific circles Callaghan was celebrated for his work on nuclear magnetic resonance; to Kiwis he was probably best known for having recently been named New Zealander of the Year. At the time he gave the speech, Callaghan was dying of cancer, and everyone who heard it realized that it was one of the last he would deliver. (He died the following month.)

"Let's get rid of the lot," Callaghan said. "Let's get rid of all the predators—all the damned mustelids, all the rats, all the possums—from the mainland.

"It's crazy," he continued, referring to his own proposal. But, he went on, "I think it might be worth a shot. I think it's our great challenge." Callaghan compared the project to the moon landing. It could be, he said, "New Zealand's Apollo program."

I listened to Callaghan's speech, via YouTube, on one of my last evenings in New Zealand. It seemed to me that what he was proposing was more like New Zealand's Manhattan Project than its Apollo program, though I could see why he hadn't framed it that way.

New Zealand's North Island is roughly 44,000 square miles. That means that it's nearly a thousand times bigger than the largest offshore island from which predators have at this point been eradicated. James Russell serves as an adviser to Predator Free New Zealand, the group that was formed to pursue Callaghan's vision. I asked him about the feasibility of scaling up by such an enormous amount. In response he showed me a graph: the size of the islands from which predators have been successfully removed has been increasing by roughly an order of magnitude each decade.

"It is a daunting scale that we're talking about," Russell told me. "But then you see the rate at which we *have* scaled up.

"Some people think scientists should just be objective," he went on. "They sit in the lab, they report their results, and that's it. But you can't separate your private life from your work life. So I do this science and then I go home and think, Wouldn't it be great if New Zealand had birds everywhere and we didn't have to worry about rats? And so that's the world I imagine."

Listening to Callaghan on YouTube also reminded me of a point that Nick Smith had made the day of the saddleback release:

in New Zealand, killing small mammals brings people together. During my travels around the country, I found that extermination, weird as it may sound, really is a grassroots affair. I met people like the Adsheads, who had decided to clear their own land, and also people like Annalily van den Broeke, who every few weeks goes out to reset traps in a park near her home in the suburbs of Auckland. In Wellington, I met a man named Kelvin Hastie, who works for a 3-D mapping company. He had divided his neighborhood into a grid and was organizing the community to get a rat trap into every 100-square-meter block. "Most of the neighbors are pretty into it," he told me.

Just about everyone I spoke to, including Hastie, expressed excitement about the latest breakthrough in extermination technology, a device designed by a company called Goodnature. And so on my last day in the country, I went to visit the company's offices. They were situated in a drab stretch of anonymous buildings not far from one of Wellington's best surfing beaches. When I arrived, I noticed a dead stoat in a plastic bag by the door. Apparently it had been killed quite recently, because it was still very much intact.

Robbie van Dam, one of the company's founders, showed me around. In the first room several young people were working at standing desks, on MacBooks propped up on cardboard boxes. A small kitchen was stocked with candy and half-empty bottles of wine. In a back room bins of plastic parts were being assembled into an L-shaped machine that resembled a portable hair dryer. Van Dam pulled out one of the finished products, known as the A24. At one end there was a CO_2 canister of the sort used in bicycle pumps. Van Dam uncapped the other end and pulled out a plastic tube. It was filled with brown goo. In the crook of the L was a hole, and in the hole there was a wire. Van Dam gingerly touched the wire, and a piston came flying out.

The A24 is designed to be screwed to a tree. The idea is that a rat, smelling the goo, which is mostly ground nuts, will stick its head into the hole, trip the wire, and be killed instantly. The rat then falls to the ground, and the device—this is the beauty part—automatically resets itself. No need to fish out rotten eggs or decaying flesh. Each CO_2 canister is supposed to be good for two dozen rats or, alternatively, stoats—hence the name. (For stoats there's different bait, made of preserved rabbit.) Van Dam also showed me a slightly larger machine, the A12, designed for possums.

"The humaneness problem was probably the hardest part," he told me. In the case of possums, it had turned out that a blow to the head wasn't enough to bring about quick death. For that reason the A12 is designed to fill the animal's skull with carbon dioxide and emulsify its brain. Goodnature also sells canisters of possum bait, laced with cinnamon. I picked up a tube. "12 out of 12 possums choose this as their final meal," the label said.

Even taking the long view—the very long view—the threat posed to New Zealand's fauna by invasive species is extraordinary. It may be unprecedented in the 80 million years that New Zealand has existed. But we live in an age of unprecedented crises. We're all aware of them, and mostly we just feel paralyzed, incapable of responding. New Zealanders aren't just wringing their hands, nor are they are telling themselves consolatory tales about the birth of "novel ecosystems." They're dividing their neighborhoods into grids and building better possum traps—ones that deliver CO_2 directly to the brain.

A couple of miles from Goodnature's headquarters is a rocky beach where little blue penguins sometimes nest. The beach is infested with rats, which can attack the nests, so Goodnature has installed some A24s along it, and van Dam took me down to have a look. It was a beautiful windy day, and the surf was high. Under the first A24, which was attached to a gnarly bush, there was nothing. Often, van Dam explained, cats or other rats drag off dead animals that have dropped from the A24, so it's not always possible to know what, if anything, has been accomplished. This had proved to be something of a sales problem.

"If people didn't find something dead under there, they were really disappointed," he said. Goodnature now offers a digital counter that attaches to the A24 and records how many times the piston has been released.

Under the second A24 there was a dead mouse; under the third and fourth, nothing. Under the fifth were two ship rats, one freshly killed and the other just a clump of matted fur with a very long tail.

AMY MAXMEN

Digging Through the World's Oldest Graveyard

FROM *Nautilus*

TWO HAMMERS, TWO shovels, four rifles. They carried their own tools. Zeresenay Alemseged, a young and driven Ethiopian fossil hunter, joined by four armed soldiers and a government official, was on a mission to the Afar Depression, a region shaped like a tornado in Ethiopia's Great Rift Valley. The Afar is bone-dry, scorching hot, and riddled with scorpions and vipers. It is regularly shaken by earthquakes and sinking deeper into the earth as the converging tectonic plates beneath it pull apart, and molten magma bubbles up through the cracks. When the magma cools, it forms sharp, basaltic blocks.

Along the road the boulders blocked Alemseged's path. He had to stop the car, lift the boulders, drive further, repeat. Dry riverbeds were smoother, but frequently the tires sank in the fine sand, and the men, sweating in the afternoon sun, pushed the Jeep onward.

Alemseged was headed to the most dangerous spot within the Afar, which even Indiana Jones types avoided because of constant conflicts between local tribes. The armed soldiers were his security. Alemseged had no salaried scientific position and refused to accompany teams led by accomplished researchers going to safer areas with fat grants. If he struck out on his own, he felt sure he could discover academic gold: ancient traces of humankind's past. This meant funding the expedition out of pocket. "I was the driver, so I didn't need to pay a driver; I was the cook, so I didn't need to pay a cook; and I was the only scientist," Alemseged said.

His aim was to explore an area called Dikika, across from a bank on the Awash River where an American paleontologist, Donald Johansen, had discovered Lucy in 1974. Her ancient skeleton's partially human, partially chimpanzee features were a clear indication of our descent from the apes. Dikika was the logical next place to look for more fossils, but no one had done so because of the risk presented by battles waged over water and land between the Afar and the Issa, pastoral tribes who inhabit Dikika. But Alemseged, who goes by Zeray (pronounced Zeh-rye), was not deterred.

Alemseged's bare-bones team reached a vast plain of sand and volcanic ashes. He knew this sediment yielded the type of fossils he was after. In December of 2000 one of the men spotted the top of a skull the size of a small orange in the dirt. Slowly, over a period of years, he and his colleagues carefully unearthed a petite skeleton of a child who had likely died in a flood and been buried in soft sand, 3.3 million years ago. She was a member of Lucy's species, *Australopithecus afarensis,* from a period about halfway between today and the time when our lineage went one way and that containing chimpanzees went the other. In 2006 Alemseged and his colleagues published their findings in *Nature.*

The child was named Selam—a word for peace in several Ethiopian languages, a wish to end the fighting in Dikika. Selam's gorillaish shoulder blades and long fingers betrayed a penchant for swinging on branches. But bones at the base of her head showed that she held it upright and therefore walked on two legs. The size of her skull suggested that her brain developed slowly through early childhood, a distinct characteristic of humans from long before modern humans evolved.

"It's the earliest child in the history of humanity," Alemseged said, enunciating each word slowly. "That discovery was 100 percent Ethiopian. It was by Ethiopians, on Ethiopian land, led by an Ethiopian scientist."

Alemseged, 45, was describing his scrappy first expedition to Dikika for me in a sparsely furnished conference room in the new facility for "antiquities research and conservation" beside the National Museum in Ethiopia's cool green capital, Addis Ababa. It was August and the facility was a hub of activity. Some of the world's foremost experts on early human evolution rushed from room to room, hurrying to collect data from their fossils before the school year started. They were here, instead of in the field, because heavy seasonal rains had flooded the dry riverbeds on which they normally drive to the Afar.

Alemseged wore two beaded bracelets on his wrist, one with the word *Ethiopia* spelled out in yellow beads on a black background. With his square jaw and confident demeanor, he looked more like a Hollywood actor playing an archeologist than the serious scientist that he is. When he's not doing fieldwork in Ethiopia, he directs the anthropology department at the California Academy of Sciences in San Francisco, where he lives with his wife and two kids.

I had come to Ethiopia in search of my own deep vision of humankind's history and fate. A flood of new discoveries coming out of the country have suggested that human traits occurred in ancient members of our tribe, the hominids, long before *Homo sapiens* entered the scene 200,000 years ago. I wanted to meet the scientists responsible for shifting our origins backward in time. Soon after I arrived, Merkeb Mekuria, an anthropologist and curator at the Ethnological Museum in Addis Ababa, greeted me. "Welcome home," he said.

Charles Darwin knew that humans evolved from apes, but he died before the strongest fossils that prove our connection with primates had been discovered. In *The Descent of Man* he wrote, "Those regions which are the most likely to afford remains connecting man with some extinct ape-like creature, have not as yet been searched by geologists."

A century later Lucy helped confirm Darwin's conjecture. By that time a vision of our origin had been born, and her skeleton was assumed to fit the story. It's one that (wrongly) persists today: apes climbed out of the trees and ambled onto their feet, dragging their fists, as the climate warmed and turned forests into grasslands. Yet it was clear to paleontologists that many more fossils were needed to test this hypothesis. That's around when Ethiopian paleontology by Ethiopians got started in earnest.

The bones of distant members of our human family are buried in tumbles of sand in Africa, and Ethiopia has unbeatable archives. The pages of human prehistory are preserved in its layers of mud, bones, and basalt. In the Afar the magma that periodically bubbles to the surface serves as a timepiece because the chemical composition of every volcanic rock betrays the stone's age. Over time the ratio of the gases trapped within it changes at a fixed, known decay rate, so you can determine whether it covered the land 4 million years ago or yesterday. Fossils located between two layers of volcanic ash and lava were left by animals that lived within that time range. Individuals belonging to Lucy's and Selam's species have been found in layers dating from almost 4 million to 3 million years ago. That means they lasted five times the duration of our own species so far. "Can we do at least as good as this primitive species?" Alemseged asked.

In 1978 a student at Addis Ababa University, located in Ethiopia's capital, was told to summarize the information on fossils discovered by Western scientists in the Afar. The student, Berhane Asfaw, had not chosen the job. It was assigned to him, as jobs were in those days, by the Derg, the communist regime that ousted the long-standing Ethiopian emperor Haile Selassie and threw Asfaw and thousands of other dissenters in jail. Of six students locked up alongside Asfaw, five were executed. Asfaw was set free.

Asfaw found solace in the geology and archeology reports he combed through. Desmond Clark, a geology professor at that time at Addis Ababa University, observed that Asfaw had done a thorough job and convinced him to pursue a graduate degree. Clark invited his mentee on an expedition to the Afar with Tim White, then a skinny, ambitious junior faculty member at Berkeley. Before the trip was over, what had begun as an assignment had become Asfaw's life passion.

"Every second I was learning," Asfaw recalled, his palms swinging upward along with his thinning eyebrows. "I had been a geology student, so I knew which rocks were old, but it was such a surprise to see fossils coming out of the sediments. It wasn't just one or two, there were plenty, and I saw hand axes and just hundreds of stone tools." Asfaw was impressed by White, now a leading expert on early human evolution. "He was so hyper, he never got tired," Asfaw said. The duo got along swimmingly. By 1981 Asfaw was off to Berkeley to finish his PhD. Soon after, his young wife followed. They appreciated Berkeley's diverse and liberal community. On Telegraph Avenue they giggled at the town's notorious bohemians.

Meanwhile, between 1983 and 1985, the Derg amplified the

devastating effects of a great drought, in which 1 million Ethiopians starved to death. Most Americans learned about the tragedy in TV ads of skeleton-thin children, and Michael Jackson's "We Are The World," and Band Aid's "Do They Know It's Christmas?" "When I saw all those people gathering to try and raise money to help the affected people, I really felt criminal to be on the outside and not doing anything," Asfaw explained. "My dream was to come back to Ethiopia and make a difference."

By 1988 the Derg's collapse was imminent, and Asfaw was eager to return home. "My plan was to survey the entire Rift Valley from north to south and look for new [fossil] sites," he said. During the transition from the Derg to the new government, the country became increasingly unstable, and there was growing conflict near the northern Eritrean border. Asfaw kept thinking about the precious hominid fossils that could be lost before they were ever found. He stressed the urgent need to preserve antiquities in grant proposals. With funds from the National Geographic Society, Asfaw organized a team including White, a Japanese friend from graduate school, Gen Suwa, and a handful of young geology, archeology, and history graduates from Addis Ababa University. By the end of the year they were off. "It was the first team with a lot of Ethiopian researchers," Asfaw said. "We were successful because we knew how to get around and which areas to avoid."

At one of the sites Suwa stumbled upon the shiny surface of a molar that was distinctly hominid. Much older than Lucy, the team called the genus *Ardipithecus ramidus,* based on the Afar words *ardi,* for "ground," and *ramidus,* for "root." They thought the species might be the first member of our family to walk on land on two legs.

Then in 1994 one of the young Ethiopians on the expedition, Yohannes Haile-Selassie—who has since become a paleoanthropologist at the Cleveland Museum of Natural History—spotted a finger bone from *Ardipithecus ramidus.* The team decided to excavate the entire region and recovered over 100 fractured pieces of a single skeleton, bones from several other individuals, and fossils of ancient animals that lived within the same period, 4.4 million years ago. That's when the real work began.

At first the team kept their fossils in the National Museum in Addis Ababa. When it overflowed, they moved them to a canary-yellow stucco building beside the museum, which had housed the

Italian government during its brief occupation of Ethiopia around 1940. There, and in an "old, moldy building" beside it, White removed hardened silt from soft bones with brushes and dental tools; Suwa took fractured pieces of the skull to Japan, where he digitally reconstructed their arrangement with a computed tomography scanner; and Asfaw compared the skull with those of ancient primates and hominids from around the world. During the course of the analysis, a skull from an older member of our ancient family was reported from Chad, but its skeleton was missing. From start to finish, the analysis took 15 years and 47 researchers to paint a full picture of *Ardipithecus ramidus*—Ardi, for short—and her surroundings. In 2009 they published 11 reports in the journal *Science.*

The following year a brand-new, five-story facility for antiquities research and conservation, funded by the Ethiopian government, opened its doors in Addis Ababa. In part this happened because of years of advocacy from Asfaw and his Ethiopian colleagues, who regularly spoke with the government about the importance of human evolution research. Grants from the United States, Japan, and France helped furnish the building and stock it with equipment. Casually called the museum facility, it abuts the old Italian government building and houses more than 250,000 ancient bones and stone tools, including 11 species of hominids—half of them discovered in the past two decades.

"Berhane deserves a lot of the credit for changing the way things are done, from the old colonial way where Westerners gained access to the countries with these resources and got publications but never invested in local scholarship," White said. "That's a lose-lose situation because the country loses and so does the science —which is done very well by folks who speak the local languages, know the geography, and understand the culture."

However, Ethiopia is still a long way from Berkeley. The electricity frequently goes out, which means Asfaw must leave his office when the sun dips below the horizon. The phone lines are dreadful, the Internet spotty. This disconnect to the rest of the world explains why Asfaw is rarely mentioned in magazine articles and books on human evolution, despite his dozen publications in top journals. He's been offered academic positions in rich countries, where he would obtain a good salary and wider recognition, but he declines. I asked why and he answered with a grin. "I am the most privileged person because I live with the fossils," he said.

With Ardi, a couple of existing views went up in smoke. Lucy's predecessor was supposed to represent an earlier step in the chain, an ape-man who hobbled through the type of savanna advertised on safari brochures. On the contrary, Ardi appears to have been a bona fide bipedal woodland dweller. Monkeys and other woodland mammals unearthed where Ardi and her kin were buried indicated that the species spent their days in the forest. Ardi's big toe remains chimpanzeelike. It's large and opposable, allowing her to climb along branches. But unlike apes, her toes are arranged in line with her foot to help her step flatly on the ground, and her pelvis is broad enough to anchor walking muscles.

"The savanna hypothesis was perfectly reasonable, until it was like, 'Oh crap, there weren't grasslands,'" said Amy Rector in the museum facility in Addis Ababa, where she was surrounded by ancient antelope skulls, their long, twisted horns extruding from wooden boxes. Rector underscored another debunked scenario: the idea of a linear progression of humankind. Two years ago the discovery of sausage-toed foot bones that match Ardi's, in layers of rock from *Australopithecus*'s time, show that a range of upright-walking species occurred simultaneously for hundreds of thousands of years.

Rector, an anthropologist at Virginia Commonwealth University, often does fieldwork in Africa and keeps her fossils in the Ethiopian museum facility. She reconstructs the context in which our ancient family members evolved by studying the animals surrounding them. "I ask myself what hominids might have seen in the area where they slept," she said. "What did they see when they woke up, what was going to eat them, where did they run to get away?"

Around 3 million years ago, Rector said, the climate appears to have warmed slightly. Some of the forests likely gave way to grasslands. But the environment as a whole was as mosaic as it is today in Ethiopia. *Australopithecus* specimens have been found around everything from woodland creatures, to grass-grazing ancestors of antelopes and gnus, to ancient hippos, crocodiles, and fish.

Back then, the sinking lowlands of the Eastern Rift Valley would still have been rather flat and fed by rivers flowing down from the mountains and the occasional land-locked lake. Walls of gray, silica-rich boulders, which formed as molten lava cooled, would have been younger than they are today, less worn by wind and rain. Ashen black mounds—created as magma ejects out of vents in the earth—would have existed also, but their location would have been different. Those seen today along the southwest corner of the Afar Depression, where three tectonic plates collide, have formed within the past several hundred years. Within a day or three, Lucy might have walked past smoldering volcanoes on this dynamic landscape, grazed on berries growing beside crocodile-infested lakes, and into the green highlands, in search of food, mates, or a safe place to rest.

During one afternoon Lucy might have come across the fresh carcass of an antelope. Famished for a meal beyond insects and roots, she might have paused to examine its succulent flesh. But at less than 4 feet tall, she would have been no match for a pack of hyenas, cackling in the distance. With a mix of hunger and fear, might she have grabbed a sharp stone and torn chunks of flesh from the beast's bones that were small enough for her to run to safety with, yet large enough to warrant the risk? After thousands of years of various individuals doing just this, might some of them have learned how to bang one rock against another and make their own sharp stones to carry?

To Alemseged, these scenarios are not far-fetched. After all, our cousins the chimpanzees stab termites with twigs, and orangutans hold leaves above their head like umbrellas when it rains. But the oldest known stone tools hail from 2.6 million years ago, long after *Australopithecus afarensis* appears to have gone extinct. Archeologists have long attributed the creation of these tools to our closer kin in the genus *Homo*.

In 2009 Alemseged realized archeologists might have been searching for stone tools with a biased image in mind. In the museum facility, he pointed to the iPhone on his desk. If you were to look for proof of telephones a century ago, he said, you'd miss them if you expected them to look like this. That year Alemseged mounted another expedition to Dikika, this time with five Jeeps and a team of 50. By then Alemseged held his current position at the California Academy of Sciences. His team was examining bones from animals in Dikika, searching for signs of action on the land when *Australopithecus afarensis* walked it. Punctures in the bones revealed that crocodiles had been voracious, and cracks whispered of antelope herds. But the causes of other scratches were unclear.

In particular a rib from a large cow and a thighbone from a small antelope bore marks that experts using electron microscopy identified as different from the rest. A sharpened stone, they said, could account for their width, shape, and angle. And radiometric dating techniques confirmed that the marks had been etched more than 3.39 million years ago, in the time of *Australopithecus afarensis.* In *Nature,* Alemseged and his colleagues reported the first signs of butchery. According to the paper, the marks are "unambiguous" evidence of stone tool use, 800,000 years earlier than when paleontologists thought it arose. "This is the first technology," Alemseged said. "It's the invention of something with the idea that it will serve some future purpose."

The finding shocked the archeology world. But Alemseged sees no reason why it should. *Australopithecus afarensis*'s humanlike hands, with long, dexterous thumbs and short fingers, would have allowed them to manipulate stones. What's more, they may have possessed the intelligence to do it. Alemseged's colleague Dean Falk, an anthropologist at Florida State University, measures the size and shape of the inside of skulls to get a sense of what ancient hominid brains looked like. Her preliminary studies on *Australopithecus afarensis* suggest that their brains were relatively advanced compared with apes' brains, with an expanded area in a front region of the brain called the prefrontal cortex—an area where intentions are processed.

With a higher-functioning brain, our ancestors may have had the cognition to create rudimentary technology. The action represents a definitive change in mental processing. "You need a plan, you need the motor skills to do it, you need to keep the task in mind for as long as it takes, and you need the motivation to go to all that work in advance of when you need the tool," Falk said. "That's all frontal-lobe stuff."

If individuals had the foresight to make stone tools, they might have also had the ability to teach one another how to do it. The transfer of complicated information among groups could signify another pivotal moment in our evolution, the origin of exceptional "social-cognitive" intelligence: the ability that builds culture among groups. Other tool-wielding animals, such as orangutans and dolphins, show a degree of social intelligence, but humans are far better at it. When given a battery of social-cognitive tests, such as producing a gesture to retrieve a reward, two-year-olds outperformed adult chimpanzees and orangutans. The children succeeded in about 74 percent of the trials, twice as often as primates.

More directly, stone tools gave our ancestors access to protein-rich food, which would have been essential to the growth of hungry big brains. Although the brain comprises just 2 percent of our body weight, it demands about 20 percent of the energy we expend each day. A bigger brain would have helped hominids build better tools and pass their knowledge on to pack members and down through the generations. It's a speculative chain of events, but the best hypothesis yet. "The emergence of stone tools is a big bang," Alemseged said. "The moment you start walking on two legs, the moment you start farming, the moment you domesticate the dog, these are major landmark moments in our history, which made us who we are today."

However, there's a lag in this chain of events if stone tool use began with *Australopithecus afarensis* some 3.4 million years ago. Drastically larger brain sizes didn't occur until about 2.5 million years ago, in our closest kin in the genus *Homo*. Shannon McPherron, an archeologist at the Max Plank Institute for Anthropology, who coauthored the report with Alemseged, said the gap might have occurred if various individuals figured out how to use stone tools independently, repeatedly over time, but never passed the knowledge on.

In this scenario, the fidelity of information transfer improved over hundreds of thousands of years. By the time *Homo* produced hand axes—oblong stones sculpted into a point, with a base that fits snugly into your palm—they were learning the craft from one another. The consistency in shape and style, as well as in abundance, is interpreted as evidence.

White is not convinced of Alemseged's early evidence of butchery. He believes crocodiles, not hominid tools, made the marks in the animal bones. Another skeptic is Sileshi Semaw, an Ethiopian archeologist at the National Center for the Investigation of Human Evolution in Spain, who codiscovered the oldest stone tools from 2.6 million years ago. White and his colleagues found signs of butchery from the same period. "Right after 2.6 million years, we have stone tools, cut marks on animal bones, the expansion of cranial capacity, and the emergence of our genus, *Homo,*" White said. "These things seem to be correlated."

Alemseged responded to the criticism by suggesting that his colleagues may be fighting to keep their stories intact. "The resistance is not based on scientific grounds," he said. In the museum facility, his team sorts through piles of rocks and bones collected in Dikika, in search of more evidence. His colleague William Kimbel, director of the Institute of Human Origins at Arizona State University, who works in the region where Lucy was found, is doing the same. With a cadre of young Ethiopian and international students now trained in the new facility, more paleontologists will be scouring Ethiopia than ever before. "Mark my words, we will find stone tools from 3.4 million years ago," Alemseged said. "I can't tell you where exactly they will be, but they will be discovered."

As I discovered during my trip to Ethiopia, the field researchers love to argue. Questioning assumptions as new evidence comes to light is, after all, the sport of science. One afternoon Alemseged, Asfaw, Suwa, and Kimbel were going at each other over a splinter of hominid skull. Suwa mangled an English idiom in an attempt to describe his objection to Kimbel's opinion; Asfaw stared at his friend of 30 years apologetically, unable to recall the phrase.

Even in their most acrimonious moments, field researchers form a tight-knit community based on respect for one another's full-body approach to science. Their colleagues in offices, who run molecular and digital analyses of fossils, may not appreciate the effort that goes into unearthing the fossils in the first place. "They don't know that the Jeep broke down in the desert and the driver fixed it on his back with an armed guard protecting him and scorpions beneath him, and he got malaria," White said. Without field research, we'd still be telling a story about how crouching apes progressed to standing man against an imaginary savanna backdrop. We'd lack the fossils to tell us that elements of humanity began millions of years ago in a mosaic of environments.

As I traveled through Ethiopia with scientists and local guides, dodging thick sheets of rain in Addis Ababa, driving past Chinese manufacturing plants outside the city and into the Afar, where I was parched, hot, and hungry, I realized just how fragile the scattered remains of our past are. They are constantly under threat by development (as African countries mine and modernize), conflict (as political situations shift), and global warming (as floods and droughts increase in severity). Ironically, our exceptional tool-making skills now threaten to lead us toward eventual demise.

When considering how long the oldest members of our family survived before they went extinct, it's impossible not to reflect on our species' fate. "When you realize that you, as an individual, are part of a very long line, you begin to take it personally, you really are afraid to cut off that line," Alemseged said to me one evening. "But I am not pessimistic, because humans are arguably the smartest species. We have the ability to reverse the damage we've done and push things forward."

SETH MNOOKIN

One of a Kind

FROM *The New Yorker*

MATT MIGHT AND Cristina Casanova met in the spring of 2002, as 20-year-old undergraduates at the Georgia Institute of Technology. Cristina was an industrial-design major with an interest in philosophy; Matt was a shy computer geek obsessed with *Star Trek*. At first Cristina took no notice of him, but the two soon became friends, and that fall they began dating. Within a year they were married.

The couple had their first child, a son, on December 9, 2007, not long after Matt completed his PhD in computer science and Cristina earned her MBA. They named him Bertrand, in honor of the British philosopher and mathematician Bertrand Russell. After a few blissful weeks, the new parents began to worry. Matt and Cristina described Bertrand to friends as being "jiggly"; his body appeared always to be in motion, as if he were lying on a bed of Jell-O. He also seemed to be in near-constant distress, and Matt's efforts to comfort him "just enraged him," Matt says. "I felt like a failure as a father." When the Mights raised their concerns with Bertrand's doctor, they were assured that his development was within normal variations. Not until Bertrand's six-month checkup did his pediatrician agree that there was cause for concern.

By then Matt had a new job, as an assistant professor at the University of Utah's School of Computing. It took two months to get Bertrand on the schedule of a developmental specialist in Salt Lake City, and the first available appointment fell on the same day as a mandatory faculty retreat. That afternoon, when Matt was able to check his phone, he saw that Cristina had left several messages. "I didn't listen to them," he told me in an e-mail. "I didn't have to. The number of them told me this was really bad."

Bertrand had brain damage—or at least that was the diagnosis until an MRI revealed that his brain was perfectly normal. After a new round of lab work was done, Bertrand's doctors concluded that he likely had a rare, inherited movement disorder called ataxia-telangiectasia. A subsequent genetic screen ruled out that diagnosis. When Bertrand was 15 months old, the Mights were told that urine screening suggested that he suffered from one of a suite of rare, often fatal diseases known as inborn errors of metabolism. During the next three months additional tests ruled out most of those ailments as well.

As Matt tried to get a foothold in his new job, Cristina struggled to care for a wheelchair-bound child whose condition seemed to worsen by the day. When Bertrand was hospitalized, she would stay by his bedside, often neglecting to eat; the constant stress contributed to osteoarthritis so severe that her doctor told her she'd need to have her right knee replaced. In April of 2009 the Mights flew to Duke University, in Durham, North Carolina, to meet with a range of specialists, including a geneticist named Vandana Shashi, whose clinical practice focuses on children with birth defects, intellectual disabilities, and developmental delays. After five days of tests and consultations, the Duke team told the Mights that there was widespread damage to Bertrand's nervous system and that some of his odd behavior—wringing his hands, grinding his teeth, staring into space—was likely due to the fact that his brain appeared to be suffering from spikes of seizurelike activity.

When Bertrand was a newborn, Matt joked to friends that he would be so relaxed as a parent that he wouldn't care which technical field his son chose to pursue for his PhD. In May of 2009 the Mights closed Bertrand's college savings accounts so that they could use the money for medical care. That fall Bertrand was rushed to the emergency room after suffering a series of life-threatening seizures. When the technicians tried to start an IV, they found Bertrand's veins so scarred from months of blood draws that they were unable to insert a needle. Later that evening, when Cristina was alone with Matt, she broke down in tears. "What have we done to our child?" she said. "How many things can we put him through?" As one obscure genetic condition after another was ruled out, the Mights began to wonder whether they would ever learn the cause of their son's agony. What if Bertrand was suffering from a disorder that was not just extremely rare but entirely unknown to science?

In September of 2012 I visited the Mights in Salt Lake City, where they lived in a two-story brick Craftsman bungalow. Matt wore a striped Brooks Brothers polo shirt and jeans; with a neatly trimmed beard and shoulder-length brownish blond hair, he brought to mind Björn Borg of the late 1970s. Cristina, who is five feet ten, with porcelain skin and long black hair, greeted me with a hug and a wry smile.

In early 2010 the couple had decided to try to have a second child. This was a gamble: if Bertrand's condition was indeed new to science, there was a chance that it was caused by a spontaneous, or de novo, mutation in the egg or sperm cell and was not in Matt's or Cristina's DNA. On the other hand, if the condition had a genetic history, the Mights could pass it on to other children. That summer Cristina learned that she was pregnant, and on April 14, 2011, she gave birth to a girl, Victoria. Within minutes of the delivery Matt and Cristina knew that their daughter was healthy; she moved with a fluidity that Bertrand never had. When I arrived at the Mights' house, Victoria was bouncing around and grabbing at her mother's sleeve. "Victoria, you need to wait for Mommy to say hello," Cristina said. To me she added, "I had no idea how easy we had it with Bertrand."

Bertrand, who was four at the time, was on the floor in the playroom, around the corner from the kitchen. He had round cheeks and a mop of brown hair. As with many children with genetic disorders, he also had some mild facial abnormalities: his eyelids drooped, and his nose was smaller than is typical, with an indentation on the bridge and slightly upturned nostrils. Two years earlier the Mights had noticed that Bertrand didn't produce tears; every time he blinked it was as if sandpaper were scraping against his corneas. To keep the resulting scar tissue from causing permanent blindness, Matt and Cristina put medicated drops and lubricating ointment in Bertrand's eyes every few hours, which made the skin around his eyes look as if it had been rubbed with Vaseline. Because Bertrand doesn't reflexively align his head with his body, his face was often pointed away from where he was trying to look, and he ground his teeth with such force that it sounded as though he were chewing on rocks. Yet the Mights told me that for all of his medical issues and his many hospitalizations, he seemed oddly immune to more ordinary ailments, such as colds and allergies.

I had brought each of the kids a small plush doll; when I placed Curious George on Bertrand's stomach, Victoria grabbed hold of Harry the Dirty Dog. When Cristina went to get something in the kitchen, she warned me not to let Victoria bite her brother. "She doesn't understand that Bertrand just can't interact with her the way everybody else can," she said. "So she gets frustrated and does everything she can to get his attention." Later, when I was lying on the floor with Bertrand and Victoria teetered into view, he seemed to flinch.

That evening, over pizza in their dining room, the Mights told me about a pattern they had noticed when Bertrand was a year old. At first, they said, he seemed to represent a challenging problem for each new specialist to solve. But as one conjecture after another was proved wrong, the specialists lost interest; many then insisted that the cause of Bertrand's illness lay in someone else's area of expertise. "There was a lot of finger-pointing," Cristina said. "It was really frustrating for us—our child hot-potatoed back and forth, nothing getting done, nothing being found out, nobody even telling us what the next step should be."

Then, in the summer of 2010, Vandana Shashi, the Duke geneticist, contacted the Mights about a new research project that was exploring whether genetic sequencing could be used to diagnose unknown conditions. There was a chance, Shashi said, that by looking for places where Bertrand's genome differed from Matt's and Cristina's, Shashi and her colleagues would be able to pinpoint the cause of Bertrand's problems. The Mights enrolled Bertrand in the study.

Genetic testing has been a part of regular medical practice since the 1970s; it enables doctors to search for mutations that cause known disorders, such as Tay-Sachs disease and sickle-cell anemia. Genetic sequencing, which entered the popular lexicon with the launch of the Human Genome Project in 1990, allows for the opposite type of search: comparing the entire genomes of people who suffer from an unknown disorder to see if they have genetic mutations in common. Sequencing also allows researchers to compare people who share genetic mutations, to see if they also share any previously unidentified disorders.

For years sequencing was too expensive for common use—in 2001 the cost of sequencing a single human genome was around $100 million. But by 2010, with the advent of new technologies, that figure had dropped by more than 99 percent, to roughly $50,000. To reduce costs further, the Duke researchers, including Shashi and a geneticist named David Goldstein, planned to sequence only the exome—the less than 2 percent of the genome that codes for proteins and gives rise to the vast majority of known genetic disorders. In a handful of isolated cases, exome sequencing had been successfully used by doctors desperate to identify the causes of mysterious, life-threatening conditions. If the technique could be shown to be more broadly effective, the Duke team might help usher in a new approach to disease discovery.

For their study, Shashi, Goldstein, and their colleagues assembled a dozen test subjects, all suffering from various undiagnosed disorders. There were nine children, two teenagers, and one adult; their symptoms included everything from spine abnormalities to severe intellectual disabilities. The researchers began by sequencing each patient and both biological parents—what's known as a parent-child trio. There are between 30 and 50 million base pairs in the human exome; the average child's exome differs from each of his parents' in roughly 15,000 spots. The researchers could dismiss most of those variations—either they corresponded to already known conditions, or they occurred frequently enough in the general population to rule out their being the cause of a rare disease, or they were involved in biological processes that were unrelated to the patient's symptoms. That left a short list of about a dozen genes for each patient.

The next step was to search through databases to see if any of those candidate genes were already associated with a rare disorder; if so, the patient was probably suffering from an unusual form of a known disease. In three of the twelve subjects, that's exactly what the researchers found. Two others had de-novo mutations on the same gene, which meant that the Duke team had likely discovered the genetic basis of a new disorder. As a diagnostic tool, sequencing seemed to work. In several of the remaining cases, the technique helped identify genetic mutations that accounted for some, but not all, of a patient's symptoms; in others it simply determined that none of the identified candidate genes were involved in the patient's illness.

Then there was Bertrand. The Duke team thought it was likely that mutations on one of his candidate genes, known as NGLY1, were responsible for his problems. Normally NGLY1 produces an enzyme that plays a crucial role in recycling cellular waste, by removing sugar molecules from damaged proteins, effectively decommissioning them. Diseases that affect the way proteins and sugar molecules interact, known as congenital disorders of glycosylation, or CDGs, are extremely rare—there are fewer than 500 cases in the United States. Since the NGLY1 gene operates in cells throughout the body, its malfunction could conceivably cause problems in a wide range of biological systems.

In September of 2011 Goldstein sent an e-mail to Hudson Freeze, a glycobiologist at Sanford-Burnham Medical Research Institute, in La Jolla, California, and the foremost authority on CDGs. Goldstein told Freeze that he believed he'd found a child suffering from a glycosylation disorder that had never before been seen. That November, Goldstein shipped Freeze a supply of Bertrand's cells. Freeze was unable to find evidence of a functioning NGLY1 gene. He soon reported back: Goldstein's hypothesis—that Bertrand suffered from a new glycosylation disorder caused by NGLY1 mutations—was almost certainly correct.

On May 3, 2012, nearly two years after the sequencing study began, the Mights met with the Duke team in an examination room of a children's hospital in Durham. Shashi explained that Bertrand's condition was probably not caused by a de-novo mutation, as the Mights had thought; rather, Matt and Cristina each had a different NGLY1 mutation, and Bertrand had inherited both. Matt and Cristina had only to look at their daughter playing on the floor to realize how lucky they'd been: Victoria had had a 25 percent chance of being born with the same disorder as Bertrand. (Later testing showed that she had not inherited either parent's NGLY1 mutation.)

Goldstein, who was meeting the Mights for the first time, spoke next. He explained that until other patients with the same condition were found, there was a chance, however remote, that Bertrand's disorder was caused by something else. Moreover, without additional cases, there was virtually no possibility of getting a pharmaceutical company to investigate the disorder, no chance of drug trials, no way even to persuade the FDA to allow Bertrand to try off-label drugs that might be beneficial. The Duke researchers estimated that there might be between 10 and 50 other patients in the country with Bertrand's condition, which would make it one of the rarest diseases in the world. "That's basically what they left us with —'You need more patients,'" Matt told me. "And I said, 'All right, we'll get more.'"

As recently as a decade ago, researchers could spend years trying to find a second case of a newly discovered disease. When a paper describing two or more cases finally appeared in one of hundreds of medical journals, it still had to be read and remembered by clinicians in order for awareness of the disorder to spread. Genetic sequencing has dramatically sped up the process, theoretically enabling a child like Bertrand to receive a tentative diagnosis in just weeks or months.

But a number of factors prevent sequencing from reaching its full diagnostic potential. As a matter of protocol, researchers typically avoid sharing test results with subjects until the research is published; the Mights didn't learn that NGLY1 was the likely cause of Bertrand's condition until months after the Duke team reached that conclusion.

Researchers also hesitate to share data with potential competitors, both to protect their funding and to insure that they get credit for their work. In their attempt to confirm Bertrand's condition, the Duke team searched for NGLY1 mutations in everyone who had been sequenced at Duke, and also combed through an exome database maintained by the National Heart, Lung, and Blood Institute. This gave them access to the genetic data of more than 6,000 people—a small fraction of those around the country who have been sequenced. Isaac Kohane, a pediatric endocrinologist at Boston Children's Hospital, told me that many researchers believe, incorrectly, that patient-privacy laws prohibit sharing useful information.

"If you want to be charitable, you can say there's just a lack of awareness" about what kind of sharing is permissible, Kohane said.

"If you want to be uncharitable, you can say that researchers use that concern about privacy as a shield by which they can actually hide their more selfish motivations."

If a team hunting for a new disease were to find a second case with the help of researchers from a competing lab, it could claim to have "solved" a new disease. But it would also have to share credit with competitors who may have done nothing more than grant access to existing data. When I asked Shashi if she could imagine a scenario that would result in one research team's publishing a paper with data from a different research group working on a similar project, she said, "Not that I can think of."

David Goldstein added, "It's not an overstatement to say that there are inherent conflicts of interest at work." Daniel MacArthur, a genetics researcher at Massachusetts General Hospital, is even more blunt. "It's an enormous deal," he told me. "And it's a big criticism of all of us, but it's a criticism we all need to hear. The current academic publication system does patients an enormous disservice."

The National Institutes of Health is taking steps to promote the sharing of genomic data about rare diseases. In January the agency awarded $9 million to researchers at Harvard Medical School to coordinate a nationwide network of centers for rare diseases. Each center will mirror the type of work being done by the NIH's Undiagnosed Diseases Program, in Bethesda, Maryland, and each will be required to share its data with the others. Kohane, who develops software to make it easier for institutions to share information, will organize the initiative. "It's creating an ecosystem that I think represents the medicine of the future," he said.

The Mights couldn't wait for the culture of scientific research to change: they had been told that Bertrand could have as little as a few months left to live. The same day that they learned about NGLY1, they began plotting ways to find more patients on their own. Several years earlier Matt had written a blog post, called "The Illustrated Guide to a PhD," that became a worldwide phenomenon; it was eventually translated into dozens of languages, including Serbian, Urdu, and Vietnamese. The popularity of the post, combined with Matt's rising profile among computer programmers, meant that almost anything he put online was quickly reposted to Hacker News, the main social news site for computer scientists and entrepreneurs. He decided to use his online presence to create what he referred to as a "Google dragnet" for new patients.

For the next three weeks Matt worked on an essay that described Bertrand's medical history in clinical detail. Matt called the result, which was more than 5,000 words long, "Hunting Down My Son's Killer," and on May 29, 2012, he posted it to his personal website. It began: "I found my son's killer. It took three years. But we did it. I should clarify one point: my son is very much alive. Yet, my wife Cristina and I have been found responsible for his death."

Half an hour after Matt hit Publish, Twitter began to light up. By the end of the day, "Hunting Down My Son's Killer" was the top story on Reddit. The next morning an editor from Gizmodo, a tech blog owned by Gawker Media, asked Matt for permission to republish the essay. In less than 24 hours the post had gone viral. The more it was shared and linked to, the higher it rose in search engines' rankings, and the easier it would be for parents of other children to find.

Eight days later the cofounder of a commercial genetic-testing company in San Francisco e-mailed the piece to a friend, Matt Wilsey. The Wilseys are one of the most prominent families in San Francisco, famous both for their philanthropic generosity and for the complicated marital life of Alfred Wilsey, Matt's grandfather, who died in 2002. Matt Wilsey, who is 36, graduated from Stanford in 2000. After working on George W. Bush's election campaign and spending five months as an aide in the Pentagon, he returned to Northern California to work as a tech entrepreneur. In the fall of 2007 he married a former classmate at Stanford. Two years later Matt and Kristen Wilsey had their first child, a girl they named Grace.

Last fall I met Matt Wilsey at the annual conference of the Society of Glycobiology, in St. Petersburg, Florida. He has a wide smile and black hair that is flecked with gray. Over lunch at an outdoor café, he told me that Grace's problems began before she was born: she was delivered by emergency cesarean section after her heartbeat dipped dangerously low. Almost immediately after Grace's birth, he and Kristen began to worry. "She just seemed out of it," Matt said. Within days Grace was admitted to the neonatal ICU. Her doctors collected a number of samples, including cerebrospinal fluid from a lumbar puncture. Three weeks later, when she was discharged from intensive care, the Wilseys still did not know what was causing their daughter's problems.

In the months to come, the Wilseys received one piece of bad news after another. They were told that Grace was not growing sufficiently, that she had low muscle tone, and that there were signs that she was suffering from developmental delays. "It was a continuous grief process," Kristen told me last December. When Grace was around six months old, the Wilseys met Gregory Enns, the director of the biochemical-genetics program at Stanford's Lucile Packard Children's Hospital. Before long Enns, whose research focuses on mitochondrial disorders, was functioning as Grace's de facto pediatrician. (Mitochondria are the so-called power plants within cells that generate most of the body's energy.)

During the next two years Matt Wilsey used his networking skills to set up meetings with specialists at institutions around the country, including Baylor College of Medicine, in Houston; the Broad Institute of MIT and Harvard; Johns Hopkins; Columbia; and the University of California, San Francisco. "We'd talk to one great doctor and say, 'Who's the best liver person in the country?'" he told me. "And then that would lead us to one person and then that person would lead us to two more. That's just kind of how we did it."

When the Wilseys first read Matt Might's blog post, it didn't occur to them that Grace and Bertrand might be suffering from the same disease. "Their phenotypes were too different," Matt Wilsey said. Grace could crawl and pull herself up to a standing position, while at age four Bertrand wasn't even able to roll over. She also had a vocabulary of more than two dozen words and was able to follow one-step directions, while Bertrand could only make indistinguishable grunts. The most striking difference, Kristen said, had to do with Bertrand's seizures. "At the time we didn't think Grace was having seizures," she said. "And so we thought, Oh, no, no—she's completely different from Bertrand. So we just ruled it out." (Later testing showed some abnormal activity in Grace's brain.)

By the spring of 2012, Grace's genome had already been sequenced twice, once at Baylor and once at Stanford. As it happened, Stanford geneticists had identified NGLY1 as a candidate gene, but they set it aside because Enns believed that Grace was suffering from an unidentified mitochondrial disorder. By the time Grace turned three that October, the Wilseys had consulted more than 100 researchers around the world, yet they were still without a diagnosis. Around this time, Kristen said, "I told Matt, 'I don't want to do this anymore. I'm just exhausted.'"

Matt asked Kristen if they could make one final trip to Baylor, and in February of 2013 the Wilseys took Grace back to Houston. They were introduced there to a young geneticist named Matthew Bainbridge. When he looked through Grace's genome, he ignored mitochondrial genes entirely—"I figured Stanford had that covered," he told me—and soon narrowed his search to three genes: one known to cause intellectual disability, one associated with a movement disorder, and NGLY1. "NGLY1 stuck out, because I'd never seen it before," Bainbridge said. When he searched a Baylor database of more than 7,000 people, he found that a handful of them had a single NGLY1 mutation but none had two.

Bainbridge next looked online for information about the gene. He quickly found "Hunting Down My Son's Killer." After reading about one of Bertrand's more unusual symptoms, Bainbridge e-mailed the Wilseys a question: Did Grace produce tears? Kristen replied almost immediately: Grace could produce tears but not very often. Then, four and a half hours later, Kristen wrote back, "After thinking about it this afternoon, it is actually very rare that Grace will make a tear. I have only seen it a handful of times in her three years." As soon as Bainbridge read that, he told me, he thought, Oh, we fucking got it.

On March 19, 2013, Bainbridge sent the Mights an e-mail. He told them that he believed he had identified a second case of Bertrand's disorder and that Matt's blog post had been instrumental in his finding it. The next day the Mights received an e-mail from Matt Wilsey. "I wanted to connect with you directly as you have heard about my daughter, Grace," Wilsey wrote. "We are so thankful to find you."

As it happened, Grace Wilsey was not the first new NGLY1 case that the Mights had uncovered. On June 3, 2012, five days after "Hunting Down My Son's Killer" was published, Joseph Gleeson, a neurogeneticist at the University of California, San Diego, e-mailed Hudson Freeze, the Sanford-Burnham glycobiologist whom the Duke team had consulted on Bertrand's case, to ask him if he'd seen the post. Gleeson told Freeze about Murat Günel, a Yale neurosurgeon and geneticist who had sequenced a pair of severely disabled siblings from Turkey, each of whom had two NGLY1 mutations. In August, Freeze confirmed that the siblings were suffering from the same condition as Bertrand.

Then, the following March, nine days before the Mights learned of Grace Wilsey, they were contacted by a researcher working with an Israeli medical geneticist named Tzipora Falik-Zaccai, who said that their group had also identified siblings with NGLY1 mutations. In May the Mights received an e-mail from Pam Stinchcomb, a woman in Georgia who had just learned that two of her daughters had NGLY1 mutations: Jordan, sixteen, who had been thought to have cerebral palsy, and Jessie, who was two. Later that month the Mights heard from a doctor in Delaware with a 20-year-old patient in whom sequencing had just revealed two NGLY1 mutations.

The most remarkable discovery came in June. Cristina Might received an e-mail from a German woman who was living in India with her husband and their severely disabled two-year-old son. (She asked that her name be withheld.) The woman had been looking online for information about how better to control her son's seizures when she came upon a blog post that Cristina Might had written about Bertrand when he was two. Within weeks the woman had sent her son's cells to Freeze, who confirmed that the boy was in all likelihood an NGLY1 patient—the first person to be identified before he had even been sequenced. Freeze told me that if someone had predicted a year earlier that the Mights would identify new patients through blog posts alone, "I'd have said, 'Ah, come on, you can't do that.'"

Thirteen months after Bertrand Might became the first NGLY1 patient in the world, the Mights had helped identify nine more cases. "There were more kids—it wasn't just our son," Cristina told me one afternoon in her kitchen. "There are parents like us, who have been lost and confused and jerked around." Matt nodded. "Even if Bertrand dies, there are kids out there that are just like him," he said.

Last November the Mights moved to a new home, in Federal Heights, an upscale Salt Lake City neighborhood at the foot of the Wasatch Range. Bertrand was about to turn six, and soon it would be difficult to carry him up and down stairs. The new house had several amenities for a family with a handicapped child, including an entrance at street level, wider hallways, and an elevator.

I visited the Mights three days after they moved. When I arrived, Cristina told me that she was two months pregnant. (Six weeks later, prenatal testing showed that the fetus had not inherited either parent's NGLY1 mutation.) Most of the family's possessions were still in boxes, but a small alcove off the kitchen had been set up as Bertrand's playroom, where, six days a week, he would spend up to three hours with a physical therapist. The room resembled an infant's nursery, with a stock of diapers, changing pads, and an assortment of soft toys. Bertrand seemed different from the way he'd been during my last visit, 14 months earlier. He had become much more expressive: he furrowed his eyebrows and scrunched his nose and, when he was pleased, grinned broadly and let out what the Mights called a "happy hoot." He was also much more coordinated, and with considerable effort he could roll over and push himself up to a sitting position. To everyone's surprise, he had even learned to communicate preferences between objects by pointing or leaning toward the one he wanted to play with.

The Mights attribute their son's improvement to several factors. Because his diagnosis revealed that Bertrand was not suffering from a seizure disorder, he was no longer on a severely restrictive diet or receiving painful, sometimes dangerous treatments such as steroids. Two over-the-counter supplements seemed to be helping as well. The first was a highly concentrated cocoa extract. "It sounds like a scam, except there's research showing that cocoa actually improves cells' energy production," Matt told me. The second was N-acetylcysteine, or NAC, an amino acid that helps produce a naturally occurring antioxidant. Bertrand hadn't been admitted to the hospital since he'd had his tonsils removed, nearly a year and a half earlier—a stark contrast to 2010 and 2011, when he'd been rushed to the hospital more than a dozen times.

In the past year the Mights and the Wilseys have formed a coalition dedicated to researching their children's condition. Patient advocacy groups have been around for decades, but it's extremely unusual for one or two families to single-handedly direct an international research agenda. It helps that both the Mights and the Wilseys have family money. (Matt Might's father is the president and CEO of Cable One, the cable television division of the former

Washington Post Company.) Since 2012 the Mights have devoted more than $100,000 a year to NGLY1-related research in Hudson Freeze's lab, while the Wilseys have spent $2 million funding researchers around the world.

Still, one of the Mights' and the Wilseys' biggest accomplishments to date required no money at all: they successfully pushed for the clinicians and researchers with whom they were working to collaborate on a single, all-encompassing clinical report on the disease. The paper, written by Gregory Enns, contained contributions from 33 authors, including Matthew Bainbridge, Hudson Freeze, David Goldstein, and Vandana Shashi. Eighteen departments from 11 institutions in the U.S., Canada, Germany, and the U.K. were represented. After the paper was all but completed, one of the challenges in getting it published was agreeing on the order in which the authors' names would appear.

Neither Murat Günel, at Yale, nor Tzipora Falik-Zaccai, in Israel, joined in the publication. Günel had been invited, but his research team had already submitted a paper elsewhere. (It was later rejected.) Freeze told me that Falik-Zaccai had stopped responding to his e-mails, and that when he inquired about the specific NGLY1 mutations of Falik-Zaccai's patients, she had all but told him that he would find out what he wanted to know when he read about it in a journal. (When I contacted Falik-Zaccai in March, she denied this account and said that she would be delighted to work with the other researchers.)

The Might and Wilsey collaboration has also prompted the NIH to study the condition. This past spring the agency began inviting NGLY1 patients to come to Bethesda for a week of tests and examinations. Bertrand was the first child to take part, and his participation has already yielded a potentially important insight. As Bertrand and the other NGLY1 patients illustrated, the inability to get rid of malfunctioning glycoproteins has devastating consequences, especially for developing infants. But Sergio Rosenzweig, an NIH immunologist who met with Bertrand, thinks it might confer one advantage as well.

Rosenzweig recently published a paper describing two siblings who rarely got viral infections despite being severely lacking in antibodies. Rosenzweig and his colleagues discovered that the siblings suffer from an ultra-rare congenital disorder of glycosylation known as CDG-IIb. The puzzle began to make sense. Since CDG-

IIb patients have trouble making glycoproteins, it is possible that the siblings had some protection from a class of viruses that are known to depend on those molecules to spread within the body. (There have been only two other known CDG-IIb patients. One is now deceased and the other does not seem to be able to ward off infection.)

When Rosenzweig learned that Bertrand rarely caught colds, he couldn't hide his excitement: perhaps Bertrand was able to avoid infections because viruses got stuck after attaching themselves to his defective glycoproteins. Rosenzweig is quick to emphasize that until he can test other NGLY1 patients he won't know if his hypothesis warrants further exploration. But if it bears out, he says, it could point to a way to treat acute infections ranging from influenza to Ebola hemorrhagic fever.

"We try to help these patients with rare diseases," he told me. "Sometimes we are able to, and sometimes we're not—but these children are teaching us a lot we didn't know about ourselves. And we can use what they are teaching us to help other people."

On a Thursday night this past February, the families of five NGLY1 patients met at the restaurant of the Estancia La Jolla Hotel and Spa, outside San Diego. They were in town for the annual Rare Disease Symposium at Sanford-Burnham, which Freeze had organized. This year's meeting was devoted entirely to NGLY1. Kristen Wilsey and Grace were the first to arrive at dinner. A few minutes later the German woman, her husband, and their son, all of whom had traveled from India, entered. Next came the Mights, followed by Kaylee Mayes, a four-year-old from Washington State who had received a diagnosis of NGLY1 disorder the previous month, and her parents, Kelsey and Daniel. Pam and Tony Stinchcomb and their youngest daughter, Jessie, arrived last. (The Stinchcombs' older daughter, Jordan, was too sick to travel.) The trip from Atlanta was just the second time that Tony Stinchcomb had been on a plane. "I'm not much into flying," Tony, a large, affable man with a gray goatee, told me. "But this seemed worth it."

That night marked the first time that two unrelated NGLY1 children were in the same room at the same time. For several hours, over ahi sliders, fish tacos, and empanadas, the families shared stories about what it had been like to care for children who were severely ill for reasons their doctors didn't understand. Just before everyone went back to their hotel rooms for the night, the German woman, fighting back tears, told me, "It feels like we've come home, but to a home we didn't know we had."

The next morning 5 researchers presented their work on NGLY1 to around 90 people, a group that included the families, a handful of their friends, and scientists from as far away as Japan. Both Matt Might's and Matt Wilsey's parents were there, as were Michael Gambello, an Emory University geneticist who had sequenced the Stinchcombs, and Gregory Enns. Cristina Might, who was five and a half months pregnant at the time, looked on from the front row. (In June she gave birth to a healthy boy.) Bertrand, who was the most severely disabled of all the children there, sat nearby in his wheelchair, humming to himself, as Kaylee amused herself on the floor.

That afternoon Matt Might gave a talk titled "Accelerating Rare Disease." After describing the effects of his blog post, he told the crowd that it was inevitable that parents of children with other newly discovered diseases would form proactive communities, much as he, Cristina, and the Wilseys had done. Vandana Shashi believes that such communities represent a new paradigm for conducting medical research. "It's kind of a shift in the scientific world that we have to recognize—that in this day of social media, dedicated, educated, and well-informed families have the ability to make a huge impact," she told me. "Gone are the days when we could just say, 'We're a cloistered community of researchers, and we alone know how to do this.'"

DENNIS OVERBYE

A Pioneer as Elusive as His Particle

FROM *The New York Times*

EDINBURGH, SCOTLAND — On October 8 last year, when the Nobel Prize in Physics was to be announced, Peter Higgs decided it would be a good day to get out of town.

Unfortunately, his car wasn't working. He got as far as lunch before a neighbor intercepted him and told him that he had won the prize.

"What prize?" he joked.

It was in 1964 that Dr. Higgs, then a 35-year-old assistant professor at the University of Edinburgh, predicted the existence of a new particle—now known as the Higgs boson, or the "God particle"—that would explain how other particles get mass. Half a century later, on July 4, 2012, he pulled out a handkerchief and wiped away a tear as he sat in a lecture hall at CERN, the European Organization for Nuclear Research in Geneva, and heard that his particle had finally been found.

Dr. Higgs, now 85, doesn't own a television or use e-mail or a cell phone. Not that he is uninterested or uninformed about the world around him. Over lunch recently he and a colleague, Alan Walker, spent half an hour parsing the implications of Scottish independence from Britain, should voters approve it Thursday. Among other things, Dr. Higgs, a longtime supporter of the Labour Party, pointed out, the departure of Scottish members would leave the British Parliament even more conservative than it is now, although he declined to say how he would vote.

But his public appearances are as rare and fleeting as the tracks of an exotic particle in the underground detectors of CERN. In a decade of covering the search for the Higgs boson, I had never managed to get a word with Dr. Higgs himself.

So I reached out to Mr. Walker, a physics professor at Edinburgh who acts as Dr. Higgs's "digital seeing-eye dog," in the words of a former student.

As a result of his bubblelike existence, Dr. Higgs doesn't really know how much commotion his award has caused, Mr. Walker said, adding, "I'm his filter."

I found Dr. Higgs looking suitably relaxed and rumpled in a lightweight parka, perusing the menu in a corner of the Cafe Royal, a downtown Victorian bistro.

Dr. Higgs said he had adjusted, sort of, to his Nobel celebrity. "I've learned to just say no," he said of people stopping him on the street and asking for a photograph.

Dr. Higgs was born in Newcastle-upon-Tyne, England, in 1929, the son of a BBC engineer and a Scottish mother. His interest in physics was tweaked when, as a schoolboy in Bristol, he realized he was attending the same school as had Paul Dirac, the British theorist who was the father of quantum field theory, which describes the forces of nature as a game of catch between force-carrying bits of energy called bosons—the same field in which Dr. Higgs would rise to fame.

Now retired from the University of Edinburgh, he lives in a fifth-floor walkup in the city's historic New Town neighborhood, around the corner from the birthplace of James Clerk Maxwell, the 19th-century Scottish theorist.

As Mr. Walker, the curator of a new show at the Royal Society of Scotland, "From Maxwell to Higgs," put it, "Scotland gave England Maxwell, and England gave Higgs to Scotland."

It was Maxwell, as a professor at King's College London, who had accomplished the first great unification of physics, showing that electricity and magnetism were different manifestations of the same force, electromagnetism, that constitutes light. It would be Dr. Higgs's fate to help physics along the next step toward a theory you could write on a T-shirt: showing that Maxwell's electromagnetism and the so-called weak force that governs radioactivity are different faces of the same thing.

As is often the case in the zigzag progress of science, however, that's not what Dr. Higgs thought he was doing. When he invented his boson in 1964, he said, "I wasn't sure it would be important."

He explained, "At the time the thought was to solve the strong force."

According to theory, bosons that carried that force, which holds nuclei together, should be massless, like the photon that transmits light. But while light crosses the universe, the strong force barely reaches across an atomic nucleus, which by quantum rules meant the particle carrying it should be almost as massive as a whole proton. Unfortunately, the theorists could not explain how the carriers of the strong force would acquire such masses.

Adapting an idea that Philip W. Anderson of Princeton had used to help explain superconductivity, Dr. Higgs suggested that space was filled with an invisible field of energy, a cosmic molasses. It would act on some particles trying to move through it, sort of like an entourage attaching itself to a celebrity trying to make it to the bar, imbuing them with what we perceive as mass.

In some situations, he noted, a bit of this field could flake off and appear as a new particle, what would come to be called the Higgs boson. Detecting this particle would be proof of the pudding that the whole spooky idea was right.

His first paper on the subject was rejected. He rewrote it, spicing it up with a new paragraph at the end, emphasizing the prediction of a new particle.

It turned out that François Englert and Robert Brout of the Free University of Brussels had beaten him into print by seven weeks with the same idea of how mass arose, although they were not as emphatic about the new particle. "They were first," said Dr. Higgs, adding that he had not known that until told as much by Yoichiro Nambu, who approved Dr. Higgs's paper for the journal that published it.

Shortly thereafter, three more physicists—Tom Kibble of Imperial College London, Carl Hagen of the University of Rochester, and Gerald Guralnik of Brown University—chimed in. Dr. Englert shared the Nobel last year with Dr. Higgs; Dr. Brout had died.

But Dr. Higgs's work turned out to be irrelevant to research on the strong force, which was ultimately transformed by the discovery of quarks.

The physicist Sheldon Glashow, now of Boston University, had proposed a theory in 1961 that unified the weak force and electromagnetic forces, but it had the same problem of explaining why the carriers of the weak part of the "electroweak force" weren't massless.

Dr. Higgs's magic field would have been just the ticket, but he and Dr. Glashow had just missed each other earlier and didn't know each other's work.

One of Dr. Higgs's duties as a beginning professor at Edinburgh was to supply daily refreshments for a Scottish summer conference in 1960. Dr. Glashow and his friends would stash wine bottles in a grandfather clock and then come back and stay up all night drinking the wine and talking about the electroweak theory while Dr. Higgs was in bed. "I didn't know they were drinking my wine," he said.

The boson became a big deal in 1967 when Steven Weinberg of the University of Texas at Austin made it the linchpin unifying the weak and electromagnetic forces. It became even bigger in 1971, when the Dutch theorist Gerard 't Hooft proved that it all made mathematical sense.

Dr. Higgs said that Benjamin Lee, a Fermilab physicist who later died in a car crash, first called it the Higgs boson at a conference around 1972, perhaps because Dr. Higgs's paper was listed first in Dr. Weinberg's paper.

The name stuck, not just to the particle but to the molasseslike field that produced it and the mechanism by which that field gave mass to other particles—somewhat to the embarrassment of Dr. Higgs and the annoyance of the other theorists.

"For a while," Dr. Higgs said, laughing, "I was calling it the 'ABEGHHKH' mechanism," reeling off the initials for all those who had contributed to the theory.

Interest in the boson, he said, has come and gone in waves. His first round of interviews came in the late 1980s, when CERN started up a new accelerator, the Large Electron-Positron Collider (LEP). There was another round when the LEP was closing down in 2000, despite claims from some scientists that they had seen traces of the Higgs boson. Dr. Higgs said he was skeptical. "They were pushing the machine beyond its limit," he said.

By then Dr. Higgs had given up doing research, concluding that high-energy particle physics had moved beyond him.

He was trying to work on a fashionable new theory called supersymmetry that would further advance the unification of forces, but "I kept making silly mistakes," he said. Indeed, he told the BBC last winter that his lack of productivity probably would have gotten him fired long ago if he had not been nominated for a Nobel Prize.

Even before the Nobel sealed his place in history, he had become an Edinburgh tourist attraction, a sort of walking monument to science, winner of the 2011 Edinburgh Award for his "outstanding contribution to the city." In 1999 he turned down an offer of knighthood but in 2014 was named a Companion of Honor by Queen Elizabeth II.

He continued to teach until he retired in 1996, but his lack of research has kept him out of the fray and the fury that has resulted from the discovery of the boson.

The Higgs boson was the final piece of what physicists have come to call the Standard Model, which sums up knowledge of the forces and particles of nature. But it is incomplete, not explaining, for example, why there is anything in the universe at all, or what dark matter and dark energy are. Moreover, lacking any evidence for supersymmetry or a more encompassing theory, physicists can't explain the mass of the Higgs itself, which standard quantum calculations suggest should be almost infinite. This has led some theorists to propose that our universe is only one in an ensemble of universes, the multiverse, in which the value of things like the Higgs is random.

Asked about that, Dr. Higgs lit up with a big grin. "I'm not a believer," he said. "It's hard enough to have a theory for one universe."

MATTHEW POWER

Blood in the Sand

FROM *Outside Magazine*

IT WAS ONLY eight o'clock on the evening of May 30, 2013, but the beach was completely dark. The moon hadn't yet risen above Playa Moín, a 15-mile-long strand of mangrove and palm on Costa Rica's Caribbean coast. A two-door Suzuki 4×4 bumped along a rough track behind the beach. The port lights of Limón, the largest town on the coast, glowed six miles away on the horizon. There was no sound except the low roar of surf and the whine of the engine straining through drifts of sand.

Riding shotgun was Jairo Mora Sandoval, a 26-year-old Costa Rican conservationist. With a flop of black hair and a scraggly beard, he wore dark clothes and a headlamp, which he used to spot leatherback sea turtle nests on the beach. Mora's friend Almudena, a 26-year-old veterinarian from Spain, was behind the wheel. The other passengers were U.S. citizens: Rachel, Katherine, and Grace, college students who had come to work at the Costa Rica Wildlife Sanctuary, a nonprofit animal-rescue center. Almudena was the resident vet, and the Americans were volunteers. By day they cared for the sanctuary's menagerie of sloths, monkeys, and birds. Working with Mora, though, meant taking the graveyard shift. He ran the sanctuary's program rescuing endangered leatherbacks, which haul their 700-plus-pound bodies onto Playa Moín each spring to lay eggs at night.

The beach's isolation made it both ideal and perilous as a nesting spot. The same blackness that attracted the turtles, which are disoriented by artificial light, provided cover for less savory human activity. In recent years the thinly populated Caribbean coast has become a haven for everything from petty theft to trafficking of Colombian cocaine and Jamaican marijuana. For decades Playa Moín has been a destination for *hueveros*—literally, "egg men"—small-time poachers who plunder sea turtle nests and sell the eggs for a dollar each as an aphrodisiac. But as crime along the Caribbean coast has risen, so has organized egg poaching, which has helped decimate the leatherback population. By most estimates, fewer than 34,000 nesting females remain worldwide.

Since 2010, Mora had been living at the sanctuary and patrolling the beach for a nonprofit organization called the Wider Caribbean Sea Turtle Conservation Network, or Widecast. His strategy was to beat the *hueveros* to the punch by gathering eggs from freshly laid nests and spiriting them to a hatchery on the sanctuary grounds. This was dangerous work. Every poacher on Moín knew Mora, and confrontations were frequent—he once jumped out of a moving truck to tackle a *huevero*.

Rachel, Grace, and Almudena had accompanied Mora on foot patrols several times over the previous weeks. (Out of concern for their safety, all four women requested that their last names not be used.) They had encountered no trouble while moving slowly on foot, but they also hadn't found many unmolested nests. On this night Mora had convinced Almudena to take her rental car. She was worried about the poachers, but she hadn't yet seen a leatherback, and Mora was persuasive. His passion was infectious, and a romance between the two had blossomed. Almudena was attracted by his boundless energy and commitment. Something about this beach gets in you, he told her.

The sand was too deep for the Suzuki, so Mora got out and walked toward the beach, disappearing in the night. Moín's primal darkness is essential to sea turtles. After hatching at night, the baby turtles navigate toward the brightest thing around: the whiteness of the breaking waves. Males spend their lives at sea, but females, guided by natal homing instincts, come ashore every two or three years to lay eggs, often to the same beaches where they hatched.

Around 10:30, Almudena got a call—Mora had found a leatherback. The women rushed to the beach, where they saw a huge female *baula* backfilling a nesting hole with its hind flippers. Mora stood nearby alongside several *hueveros*. One was instantly recognizable, a 36-year-old man named Maximiliano Gutierrez. With his beard and long reddish brown dreadlocks, "Guti" was a familiar presence on Moín.

Mora had forged a reluctant arrangement with Guti and a few other regular poachers: if they arrived at a nest simultaneously, they'd split the eggs. After measuring the turtle—it was nearly six feet long—Mora and Rachel took half the nest, about 40 cue-ball-sized eggs, and put them into a plastic bag. Then Guti wandered off, and the turtle pulled itself back toward the surf.

When they returned to the road, a police patrol pulled up. The cops warned Mora that they had run into some rough characters earlier that night, then drove off as Mora and the women headed south, toward the sanctuary, just six miles away. Soon they came upon a palm trunk laid across the narrow track—a trick the *hueveros* often played to mess with police patrols. Mora hopped out, hefting the log out of the way as Almudena drove past. Just as Mora put the log back, five men stepped out of the darkness. Bandannas covered their faces. They shouted at everyone to put their hands up and their heads down. Then they grabbed Mora.

"Dude, I'm from Moín!" he protested, but the men threw him to the ground.

Masked faces crowded into Almudena's window. The men demanded money, jewelry, phones, car keys. They pulled Almudena out and frisked her, and the Americans stayed in the car as the men rifled through it, snatching everything of value, including the turtle eggs. Almudena saw two of the men stuffing a limp Mora into the tiny cargo area. The four women were jammed into the back seat with a masked man sprawled on top of them. As the driver turned the Suzuki around, Almudena reached behind the seat and felt Mora slip his palm into hers. He squeezed hard.

The driver pulled off next to a shack in the jungle, and the men, claiming to be looking for cell phones, told the girls to lift their shirts and drop their pants. Mosquitoes swarmed them. After being frisked, Almudena caught a glimpse of two of the men driving off in the Suzuki. Mora was still in the trunk.

The four young women sat on logs behind the hut with two of their captors. The guys seemed young, not more than 20, and were oddly talkative for criminals. They said they understood what the conservationists were trying to do, but they needed to feed their families. One said that Mora "didn't respect the rules of the beach."

The men announced that they were going to get some coconuts, walked away, and never came back. After an hour the women decided to make a break for it. Huddled close together, they walked down to the beach and headed south toward the sanctuary. They were terrified and stunned, barely speaking and moving on autopilot. Two hours later they finally reached the gate but found no sign of Mora. Almudena started to sob. A caretaker called the police in Limón, and soon a line of vehicles raced north along the beach track. At 6:30 a.m. the police radio crackled. They had found Almudena's car, buried up to its axles in sand. There was a body beside it.

Mora was found naked and facedown on the beach, his hands bound behind him and a large gash on the back of his head. The official cause of death was asphyxiation—he'd aspirated sand deep into his lungs.

The news spread quickly. A chorus of tweets cast Mora as an environmental martyr akin to Chico Mendes, the Brazilian rainforest activist who was assassinated in 1988. The BBC, the *New York Times*, and the *Washington Post* picked up the story. An online petition started by the nonprofit Sea Turtle Restoration Project called on Costa Rican president Laura Chinchilla for justice and gathered 120,000 signatures. Paul Watson, the founder of the Sea Shepherd Society and the star of *Whale Wars*, offered $30,000 to anyone who could identify the killers. "Jairo is no longer simply a murder statistic," Watson wrote. "He is now an icon."

There was a sense too that this killing would be bad for business. Long the self-styled ecotourism capital of the world, Costa Rica relies on international travelers for 10 percent of its GDP. "What would have happened if the young female North American volunteers were murdered?" wrote one hotel owner in an open e-mail to the country's ecotourism community. "Costa Rica would have a huge, long-lasting PR problem." Not long after, President Chinchilla took to Twitter to vow that there would be "no impunity" and that the killers would be caught.

That task fell to detectives from the Office of Judicial Investigation (OIJ), Costa Rica's equivalent of the FBI, and Limón's police department. The OIJ attempted to trace the victims' stolen cell phones, but the devices appeared to have been switched off and their SIM cards removed. Almudena, Grace, Katherine, and Rachel gave depositions before leaving the country, but it was clear that finding other witnesses would be a challenge.

Moín is backed by a scattering of run-down houses behind high walls. It's the kind of place where neighbors know one another's business but don't talk about it, especially to cops. The *hueveros* met OIJ investigators with silence. When detectives interviewed Guti, he was so drunk he could barely speak.

Not everyone kept quiet, though. Following the murder, Vanessa Lizano, the founder of the Costa Rica Wildlife Sanctuary, dedicated herself to fighting for her fallen colleague's legacy. I e-mailed her and asked if I could come visit, and she welcomed me.

I flew to San José two weeks after the killing, arriving at the sanctuary after dusk. Lizano, 36, unlocked a high gate adorned with a brightly painted butterfly. "Welcome to Moín," she said in a theatrical voice, her auburn hair pulled back in a ponytail. The property covered about a dozen acres of rainforest and was dotted with animal pens. Paintings of Costa Rica's fauna adorned every surface. Lizano opened a pen and picked up a baby howler monkey, which wrapped its tail around her neck like a boa. "I keep expecting Jairo to just show up," she said. "I guess I haven't realized it yet."

Lizano had been running a modeling agency in San José in 2005 when she and her parents decided to open a butterfly farm near the beach. She leased a small piece of land and moved to Moín with her infant son, Federico, or Fedé, her parents, and a three-toed sloth named Buda. They gradually transformed the farm into a sanctuary, acquiring rescued sloths and monkeys, a one-winged owl, and a pair of scarlet macaws seized from an imprisoned narcotrafficker. Fedé pulled baby armadillos around in his Tonka trucks and shared his bed with Buda.

Lizano operated the sanctuary with her mother, Marielos, and a rotation of international volunteers, who paid $100 a week for room and board—a common model for small-scale ecotourism in Costa Rica. The sanctuary was never a moneymaker, but Lizano loved working with the animals.

Then, one day in 2009, she discovered several dead leatherbacks on the beach that had been gutted for their egg sacs. "I went crazy," she says. She attended a sea turtle conservation training program in Gandoca run by Widecast, a nonprofit that operates in 43 countries. There she met Mora, who'd been working with Widecast since he was 15. Lizano arranged for the organization to operate a turtle program out of her sanctuary, and in 2010 Mora moved to Moín to help run it.

They soon developed something like a sibling rivalry. They'd psych themselves up by watching *Whale Wars,* then compete to see who could gather more nests. Normally a goofball and unabashed flirt, Mora turned gravely serious when on patrol. He loved the turtles deeply, but he seemed to love the fight for them even more. Lizano worried that his stubbornness may have made things worse on the night he was killed.

"Jairo wouldn't have gone without a fight," she said. "He was a very, very tough guy."

Lizano told me that her mission was now to realize Mora's vision of preserving Playa Moín as a national park. She had been advocating for the preserve to anyone who would listen—law enforcement, the government, the media. It was a frustrating campaign. The turtle program had been shut down in the wake of the killing, and poaching had continued. Meanwhile Lizano seemed certain that people around Moín knew who the killers were, but she had little faith in the police. On the night of the murder, when Erick Calderón, Limón's chief of police, called to inform her that Mora had been killed, she screamed at him. Since 2010, Calderón had intermittently provided police escorts for the sanctuary's patrollers, and by 2013 he'd suspended them because of limited resources. Prior to the killings, Lizano and Mora had asked repeatedly for protection, to no avail. The murder, Lizano said, was Calderón's fault.

But there was plenty of recrimination to go around. The ecotourism community blamed Lizano and Widecast for putting volunteers at risk. The family of one of the Americans, Grace, had demanded that Widecast reimburse her for her stolen camera, phone, and sneakers. Lizano told me the accusations were unfair. "The volunteers knew what they were getting into," she said. "We would say, 'It's up to you if you want to go out.'"

Still, she was overwhelmed with guilt. "I know Jairo was scared, because I used to tease him," she said. "We'd make fun of each other for being afraid. We'd always kid around that we would die on the beach." She'd tell him that she wanted her ashes carried into the surf by a sea turtle. Mora was less sentimental. "He always said, 'You can do whatever, I really don't care. Just drink a lot. Throw a party.'"

We sat in the open-air kitchen, and Lizano held her head in her hands. "If you've got to blame somebody, blame me," she said. "I was the one who took Jairo and showed him the beach, and he fell in love."

Mora was born in Gandoca, a tiny Caribbean town near the Panama border. He caught the wildlife bug early, from his grandfather, Jerónimo Matute, an environmentalist who helped found the Gandoca-Manzanillo Wildlife Refuge, a sea turtle nesting area. Jairo began releasing hatchlings at age six. Once he became a full-time Widecast employee, he sent much of his salary home every month to his mother, Fernanda, and completed high school through a correspondence program.

By 2010 Mora had moved to Moín, living in a tiny room over the sanctuary's kitchen. Some days Mora and the volunteers—college students, mainly, from all over the world—counted poached nests or monitored the sanctuary's hatchery; some nights they'd go on patrol. Mora was clear about the risks involved, and some chose not to go, but others joined eagerly. It didn't seem that dangerous, especially in the early days, when the Limón police accompanied the patrols.

Still, there were tensions from the beginning. During nesting season the *hueveros* squatted in shacks in the jungle. Most were desperately poor, many were addicts, and all considered Lizano and Mora competition. Lizano had no qualms about reporting poachers to the police.

A leatherback typically lays 80 fertilized eggs and covers them with about 30 yolkless ones. Poachers consider the yolkless eggs worthless and usually toss them aside. Lizano and Mora often placed those eggs on top of broken glass, causing a poacher to cut himself while digging for the good ones. Lizano even set volunteers to work smashing glass to carry in buckets to the beach. She sometimes found obscene notes scrawled in the sand. She'd write back: *Fuck You.*

Lizano got caught in shootouts between police and poachers at the beach four times, once having to duck for cover behind a leatherback. In April 2011 she was driving alone at night on Moín when she came across a tree blocking the road. Two men with machetes jumped out of the forest and ran toward her truck. She floored it in reverse down the dirt road, watching as the men with the machetes chased, their eyes full of hate.

In the spring of 2012, Calderón suspended the police escorts. Limón had the highest crime rate in Costa Rica, and the police chief was spread too thin trying to protect the city's human population, never mind the turtles. Mora and Lizano shifted to more conciliatory tactics. They hired ten *hueveros* and paid each of them a salary of $300 per month, using money from the volunteers' fees. In return the men would give up poaching and work on conservation. Guti was one of the first to sign on. The *hueveros* walked the beach with the volunteers, gathering nests and bringing them to the hatchery. It was a steep pay cut—an industrious *huevero* can make as much as $200 a night—so Lizano pushed the idea that the poachers could eventually work in the more viable long game of ecotourism, guiding tourists to nesting sites. But the money for the project quickly ran out, and Lizano wasn't surprised when poaching increased soon after.

Around the same time, a menacing poaching gang showed up on Playa Moín. They seemed far more organized than the typical booze-addled *hueveros.* The group dropped men along the beach by van, using cell phones to warn each other of approaching police. They were led by a Nicaraguan named Felipe "Renco" Arauz, now 38, who had a long criminal history, including drug trafficking and kidnapping.

In April 2012 a group of men armed with AK-47's broke into the hatchery, tied up five volunteers, and beat a cousin of Mora's with their rifle butts. Then they stole all 1,500 of the eggs that had been collected that season. Mora, out patrolling the beach, returned to find the volunteers tied up. He went ballistic, punching the walls. Then he exacted vengeance, going on a frenzy of egg gathering, accompanied once again by armed police protection. Mora collected 19 nests in three nights, completely replacing the eggs that had been stolen. But a few weeks later Calderón once again suspended escorts, and no arrests were made.

A month after the hatchery raid, in May 2012, the dangers became too much even for Lizano. She was at a restaurant in downtown Limón when she spotted a man taking Fedé's photo with his cell phone. She recognized him as a *huevero* and confronted him angrily: "It's me you want. Leave the kid out of it." The man laughed at her. That was the final straw. She moved with Fedé back to San José, returning to Moín alone on weekends.

Mora remained, however, and when the 2013 season began in March, he returned to his patrols—mostly alone, but occasionally with volunteers. By this point the volunteer program was entirely Mora's operation. The Americans, who arrived in April, knew there were risks. But according to Rachel, Mora never told her about the raid on the hatchery the year before. She entrusted her safety to him completely. "I had gone out numerous times with Jairo and never really felt in danger," she told me. "I knew he was there and wouldn't let anything happen to me."

But just a few weeks before his death, Mora told a newspaper reporter that threats were increasing and the police were ignoring Widecast's pleas for help. He called his mother, Fernanda, every night before he went on patrol, asking for her blessing. When Lizano saw Fernanda at Mora's funeral, she asked for her forgiveness.

"Sweetie," Fernanda replied, "Jairo wanted to be there. It was his thing."

Click-click.

The cop next to me, young and jumpy in the darkness, pulled his M4's slide back, racking a cartridge. As I crouched down, I saw two green dots floating—the glow-in-the-dark sights of a drawn 9 mm. About 100 yards off, the police had spotted a couple of shadowy figures. *Hueveros.*

I was on patrol. Following Mora's killing, the sea turtle volunteer program had been suspended, but two of Mora's young protégés, Roger Sanchez and his girlfriend, Marjorie Balfodano, still walked the beach every night with police at their side. Sanchez, 18, and Balfodano, 20, were both diminutive students, standing in bare feet with headlamps on. They weren't much to intimidate a poacher, but Sanchez was fearless. Before we set out he told me with earnest bravado that he planned to patrol Moín for the rest of his life. When we saw the *hueveros,* we'd been walking for three hours alongside an escort of five officers from Limón's Fuerza Pública, kitted out with bulletproof vests, sidearms, and M4 carbines. Perhaps it was just a publicity stunt by Calderón, but it was a comforting one. We had encountered a dozen plundered nests, each one a shallow pit littered with broken shells. The *hueveros,* it had seemed, were just steps ahead of us.

Then the cop on my right noticed two figures and pulled his gun. Three of the police told us to wait and confronted the two men. After several minutes we approached. The cops shone their flashlights on the poachers and made them turn out their pockets. One wore a knit cap, and the other had long reddish dreadlocks —Guti. They were both slurry with drink, and the cops seemed to be making a show of frisking them. The men had no contraband, so the cops let them stumble off along the beach.

After a while the radio crackled. Another police truck had found two nesting leatherbacks. We rushed to the spot. In the darkness, a hump the size of an overturned kiddie pool slowly shifted in the sand. The *baula*'s great watery eyes looked sidelong toward the sea as it excavated a nest in the beach with back flippers as dexterous as socked hands. With each labored effort it delicately lifted a tiny scoop of sand and cupped it to the edge of the hole. Sanchez held a plastic bag in anticipation, ready for her to drop her clutch.

Then Guti's drunken companion stumbled up to us, knelt beside Sanchez, and offered a boozy disquisition on sea turtle biology. The cops ignored him, and the spooked animal heaved forward, dragging her bulk away without laying any eggs. A few more heaves and the foaming waves broke over the turtle's ridged carapace.

The night wasn't a complete loss, though. A short distance away, the second leatherback had laid its nest. Soon a second patrol truck pulled up and handed Sanchez a bag of 60 eggs. We hitched a ride back to the sanctuary and a wooden shed packed with Styrofoam coolers. Sanchez opened one, sifted beach sand into the bottom, then began placing the eggs inside. I noticed that a pen had been stuck into one of the coolers. Next to it a set of stylized initials was scratched into the Styrofoam: *JMS.* Altogether there were perhaps 1,000 eggs in the coolers. Almost all of them had been gathered by Mora.

A couple of days later I went to see Erick Calderón at the police headquarters in Limón. With his small build and boyish face, he seemed an unlikely enforcer, and he'd clearly been affected by the pressure the killing had brought on his department. Since the murder, Calderón said, the police had patrolled Moín every night. "I want to make the beach a safer place, control poaching of eggs, and educate the population so the demand isn't there," he said. But it was unclear how long he could sustain the effort. He said that only a dedicated ecological police force would make a lasting impact. They'd need a permanent outpost on Moín, a dozen officers supplied with 4×4s and night-vision goggles.

Then Calderón insisted that Mora's murder was an anomaly and that Costa Rica was "not a violent society"—an assertion belied by the fact that the previous afternoon a shootout between rival gangs had happened just a few blocks from the station. He seemed ashamed that the murder had happened on his watch, that Lizano had screamed at him. "I know Jairo was a good guy," he told me.

That afternoon I met up with Lizano's father, Bernie. His means of processing his sorrow had been to turn himself into a pro bono private investigator. A former tuna fisherman, Bernie was 65, with a full head of white hair and a pronounced limp from an old boating accident. As we drove around Limón, he seemed to know everyone's racket, from the drug kingpins behind razor-wire-topped fences to a guy on a corner selling drinks from a cooler. "He keeps the turtle eggs in his truck," Bernie whispered conspiratorially. At one house he stopped to chat with a shirtless, heavily tattooed man. The guy offered his condolences, then said, "Let me know if you need any maintenance work done." As Bernie pulled away he chuckled: "*Maintenance.* That guy's a hit man."

We drove to a squat concrete building with dark-tinted windows on the edge of town—the office of the OIJ. After Bernie and I passed through a metal detector, one of the case's detectives, tall and athletic, with a 9 mm holstered in his jeans, agreed to speak with me anonymously. He said that OIJ investigators in Limón were the busiest in the country due to drug-related crime. I asked whether he thought the killers were traffickers, and he shook his head wearily. "If they were narcos, it would have been a disaster," he said. "Every one of them would have been killed."

Like Calderón, he promised that Mora would not be a mere statistic. He insisted that they were closing in on serious leads. Walking out, Bernie told me he had spoken in private with the detective, to whom he'd been feeding every scrap of information he'd gotten. "He told me, 'We are very close to getting them, but we don't want them to know because they'll get away.'"

Bernie's PI trail led back to Moín, where he had tracked down a potential witness—a man who lived near the beach. The man had been the first to find Mora early on the morning of May 31. He walked Bernie to the spot where he'd found the body. As he described it, there were signs of a struggle from the footprints around the car. It looked to him like Mora had escaped his captors and dashed down the beach. Another set of tracks seemed to show a body being dragged back to the vehicle.

Bernie had begged the man for some clue, mentioning Paul Watson's reward, which had now swelled to $56,000. "He said, 'No, no, I don't need the money. It's not that I don't need it, it's just that they did something very bad.'" If he talked, he was sure that he and his family would be killed.

On July 31 the OIJ conducted a predawn raid, called Operation Baula, at several houses around Limón. Dozens of armed agents arrested six men, including Felipe Arauz, the 38-year-old Nicaraguan immigrant suspected of being the ringleader of the violent *hueveros.* A seventh man was caught 10 days later. The suspects were Darwin and Donald Salmón Meléndez, William Delgado Loaiza, Héctor Cash Lopez, Enrique Centeno Rivas, and Bryan Quesada Cubillo. While Lizano knew of the alleged killers, she was relieved that she hadn't worked with them. "Thank God none were my poachers," she said.

Detectives from the OIJ had been talking to informants and quietly tracking Mora's stolen cell phone. According to court documents, one of the suspects, Quesada, 20, had continued to use it, sending incriminating texts. One read: "We dragged him on the beach behind Felipe's car and you know it."

To Lizano the motive was clearly revenge, but the authorities cast the crime as "a simple robbery and assault." They also laid blame on Mora and Lizano's failed attempt to hire poachers for conservation. An OIJ spokesman claimed that the program had bred resentment among *hueveros.* The accusation infuriated Lizano. "They're just looking for a scapegoat," she said.

Lizano thought that the authorities were deflecting blame. It turned out that on the night of the murder, a police patrol had encountered several of the suspects—they were the same men the cops had warned Mora about. A few hours later the gang lay in wait. Whether or not they intended to kill Mora will be argued at the trial later this spring.

Even so, the arrests haven't brought much closure to those closest to Mora. Almudena, back in Madrid, was deeply depressed when I reached her. "Jairo is dead," she said. "For me there is no justice." The only positive outcome, as she saw it, would be for a preserved beach. "In ten years there have to be turtles at Moín," she said. "If not, this has happened for nothing."

Lizano, meanwhile, redoubled her efforts to protect Moín. Any legislative change to preserve the beach is far off, and the turtles now face an additional threat—a massive container-port development project that a Dutch conglomerate hopes to build nearby. Still, Lizano told me, "I really believe it has to continue. I can't stop and let the poachers win. For me it's not an option."

In July, Lizano brought Fedé back to Moín. She woke him up one morning before sunrise, and together with a group of volunteers they walked to the beach. The night before, at the sanctuary, the first turtle hatchlings had broken up through the sand in their Styrofoam-cooler nests. Lizano showed Fedé how to lift the tiny flapping things out and set them gently on the sand. The people stood back and watched as the turtles inched down the beach, making their way toward the breaking waves and an uncertain future.

SARAH SCHWEITZER

Chasing Bayla

FROM *The Boston Globe*

"THIRTY METERS," MICHAEL MOORE called out.

Moore braced himself against the steel of the Zodiac's platform tower as the boat closed in on the whale in the heaving Florida waters. Through the rangefinder he could see the tangled mass of ropes cinched tightly around her. It was impossible to tell where the ropes began and where they ended.

This much he knew. The ropes were carving into her. Bayla was in pain.

He was tempted to look away. It was almost too much to see.

Her V-shaped spray erupted, then disappeared into a mist as she slipped beneath the surface. A spot plane circling overhead radioed. They could still see her silhouette. She hadn't gone deep.

"Get in close if you can," Moore said to the boat's driver.

Bayla would come up for air again soon.

Then he would have his chance.

For nearly three decades Moore had dedicated himself to North Atlantic right whales like Bayla. He knew every inch of their anatomy, every detail of the strange and glorious physiology that made them so astoundingly powerful and so utterly defenseless against the ropes.

They were majestic and doomed, his love and his burden. He had believed he could save them. But in those 30 years he'd watched too many succumb. Saving just two female whales a year could stabilize a population that humans had driven down to just 450 from the teeming thousands that once greeted settlers to the New World.

And so he had raced down the interstate through a driving New England snowstorm after the e-mail had come.

The details were grim. Bayla had been spotted off the coast of Florida three weeks earlier, on Christmas Day, 2010. Rope anchored in her mouth. It coiled around her flippers in a skein of tangled loops. With every move it pulled tighter.

The rope was likely polypropylene, a synthetic weave favored by Northeast fishermen and lobstermen for its brute strength against the abrading forces of a rocky-bottomed seabed. Blubber was no match for it. Bayla's body was cut open in places, as though by cheese wire.

Her back sloped alarmingly, a sign of emaciation from hauling rope more than 10 times her length, possibly for months. It was like she had been swimming with an open parachute.

Biologists from Florida and Georgia had tried to cut the ropes. But Bayla threw them off with heaves of her massive tail and stunningly quick hairpin turns. They tried again a day later. Still they couldn't get near enough. She was a bucking bronco.

And so they summoned Moore.

Moore had engineered something that could be a breakthrough for rescuers, a way to sedate whales at sea. The man standing to his left on the Zodiac platform held the instrument Moore had conceived for the task: a pressurized rifle tipped with a dart and syringe filled with 60 cc's of a sedative so powerful that a few drops on human skin could kill.

Bayla was probably seven tons, but you can't weigh a free-swimming whale. If the estimate was wrong, an overdose could plunge Bayla into a catastrophic slumber and she would drown.

Moore scanned the horizon. Fishing charters and Disney Cruise liners jockeyed for space at the shore. Ahead, the vast reach of the Atlantic met at every point with the prickling Florida sun.

He knew that the work of a lifetime shouldn't come down to a single moment. He was the father of four grown boys. He loved his wife. His home was an island in Marion Harbor. He had published scores of peer-reviewed papers and commanded millions in grant money.

Yet the vow he had made to himself as a young man, the thing he had dedicated his career and heart to, remained unfulfilled. For Moore, nearing retirement and running out of ideas, there might be no more chances.

Blow spouted off the port bow.

"Twenty-one meters," he called to the man with the dart rifle.

Bayla's hobbled body arced through the swells.

"Shoot."

For more than a thousand years, humans hunted the North Atlantic right whale. Big, slow, and without guile, the whales often ventured up to boats, rolled over, and eyed their pursuers with peering curiosity, making for easy marks. Endowed with abundant blubber, right whales also floated after being killed. It was a grimly convenient attribute that, legend has it, afforded them their name. They were the right whale to kill.

Basques hunted them in the Dark Ages. The rest of the European continent followed. Pilgrims on the *Mayflower* spied right whales as they came into Cape Cod Bay, a feeding ground for the vast animals. "Every day we saw whales playing hard by us," one passenger wrote. The ship's master and mate lamented they didn't have the tools to kill the whales. They soon would. An industry quickly took root in maritime New England. On a single day in January of 1700, colonists killed 29 right whales off the Cape.

Oil from right whale blubber helped propel the Colonial economy, lighting homes and stores and creating wealth and prosperity. By the time whale oil demand faded and right whales were protected from hunting, in 1935, their numbers had been reduced from the thousands to some 100 in the North Atlantic.

Today right whales remain among the rarest animals on earth. Their pursuers are whale-watching boats and a legion of scientists who track them in the hope of figuring out why their numbers hover stubbornly in the low hundreds, a population so fragile that it could be wiped out with one algal bloom.

Researchers say the peril can be traced once again to humans —this time because right whales get in our way, or we in theirs. Dubbed the "urban whale," North Atlantic rights live along the eastern seaboard, one of the most developed coastal zones in the world. Migrating from southern calving grounds to northern feeding climes is an industrial obstacle course for the whales, studded with pollution, noise, ships, and, most devastatingly, fishing gear —often buoy-tethered ropes leading to lobster pots and crab traps.

Among whales, rights are particularly prone to getting caught in the gear, with 83 percent of those tracked by scientists bearing scars from entanglement. Their special susceptibility, researchers say, owes to feeding with open mouths—filtering tiny prey through their long plates of baleen but also taking in the ropes so common in their domain. Once snagged, the whales frantically spin their bodies trying to get free, but their gyrations instead loop the ropes around flippers and flukes. Unlike weaker whale species, which tend to drown when entangled, right whales, which run to 50 feet and 60 tons or more, often have the strength to swim with the lacerating ropes for months, sometimes years.

It's not known how many right whales die from entanglement. Scientists have recorded an average of four such confirmed and presumed deaths per year since 2008, but they believe many more perish this way unrecorded. In a species plagued by abnormally low reproductive rates, in some years with a single calf born in the known population, scientists worry that deaths from ropes could be right whales' ultimate undoing—Moore chief among them.

In the summer of 1979, a grungy 28-foot sailboat with a scruffy crew docked in Newfoundland's treeless peninsula of Bay de Verde, an outcropping of houses, a fish processing plant, and a bar called the Holding Ground.

Word went around that the boat's inhabitants were long-hairs, American college students fired up by the "Save the Whales" movement who had hitchhiked to Newfoundland to study humpback whales. The students played sea shanties on concertina and guitar. They drank Dominion Ale at the Holding Ground and tried to convince fishermen that whales ensnared in cod nets were not nuisances but wonders.

Soon a 23-year-old British veterinary student joined them.

Michael Moore was on expedition, as students' research journeys to remote spots were called at the University of Cambridge. He wasn't certain he wanted to spend his life tending to dogs and cats, as he had assumed he would. Studying whales was a year-long diversion as he wrestled with his career plans.

To the Americans, Moore projected quirky English certitude. He had graduated from Winchester College, an elite boarding school, and was now at Cambridge. He took his tea every day at 4 p.m. He read Thomas Hardy aloud. On his first night aboard the sailboat, as the Americans climbed into their berths in salted dungarees and cable-knit sweaters, Moore opened a leather satchel. He pulled out a pair of striped pajamas so crisp they might still have had Harrods tags attached.

In the mornings Moore and his boat mates woke to the frenzy of gulls feasting on cod stomachs gutted by fishermen after predawn hauls. They pulled on oilskins and ventured into the cold emerald bay with the boat's hand-cranked engine belching diesel fumes. They chugged around the peninsula for hours watching humpbacks lunge at schools of capelin.

Moore dutifully jotted observations about the whales in his journal and said little to the Americans. They ribbed him for his punctually taken Earl Grey. One afternoon he returned the volley with dry, cordial wit. The Americans were taken aback. Moore had found his way. "Beginning to relax and feel part of the machine," he wrote in his journal. "There are some real super people around here—all fine and kind and loving."

From the time of his childhood, Moore had felt somewhat apart. He'd been 12 when his mother told him that his father was manic-depressive. The news stunned him. He'd always thought of his father as a steady rock, a country doctor content with his practice. He'd had no idea that his father was undergoing a grueling course of convulsive shock therapy or that his mother had come upon him after an attempted suicide. The family dynamic now became clear to him: his mother tended his father and his father tended his patients. Alone many afternoons, he wandered from his home down to the railroad tracks, where he kept tabs on a badger family, an experience that steered him to study the animal condition.

Now at last he was on the inside.

His descriptions of the whales grew animated and lively. They were grand, clever, and powerful, the sea's benign emperors, yet astonishingly vulnerable. One day the researchers came upon a humpback caught in a cod trap under the cliffs. The whale was thrashing in panic. Other humpbacks were circling helplessly. "I need to go in and cut the net," Moore said as he and the others watched in horror. It would have been reckless; the nets could readily have entangled him too. He was about to dive when other humpbacks helped the whale break free, leaving Moore's declaration flapping like the flag of an impetuous explorer who had stumbled on something but wasn't yet sure what it was.

Moore stayed on with the researchers at summer's end and sailed with them to the humpbacks' wintering grounds in the Caribbean. As Christmas approached and carols played on Radio Antigua, he thought of his mother. She had died two years earlier. He thought about how burdened she had been by his father's illness and about the guilt he'd felt for being unable to lighten her load.

One night Moore woke in his berth. He had been dreaming of whales swimming around the coral reef where the boat had anchored. The whales were singing in the dream. As consciousness pushed aside the dream, Moore realized the calls and whistles of the whales were not the stuff of his mind but real and coming through the hull, a chorus of longing and kinship.

Surrounded by the sounds, Moore realized he would spend his life studying, helping, and learning from these creatures.

Bayla was Picasso's new beginning.

Bayla's mother had lost a calf in the fall of 2007. It was her first, and it lived only a few months. Researchers don't know where or why the calf died, but they assumed something catastrophic had happened when Picasso appeared without it. No right whale mother would have abandoned her young before a year spent together.

For Picasso, the death of the baby had to have been wrenching. Right whale mothers are known to swim frantically for days after the death of a calf, searching for the little one no longer at their side.

Researchers spotted Bayla for the first time on January 2, 2009, swimming alongside her mother and a pod of bottlenose dolphins off the coast of Georgia.

Like all right whale encounters, the sighting of Picasso with her days-old calf was a matter of luck. In an era when wild animals are routinely monitored, their every movement documented for years running, whales are the exception. Whales can't be banded like birds or collared like a wolf, and implanted tags can fail after a short time. The legions of scientists who study them often can only guess their location in the depths.

Researchers catalog sightings in a database painstakingly maintained by the New England Aquarium since 1980. The database distinguishes the whales often by cream-colored skin patches that grow in the same spots where human hair sprouts—on heads, above eyes, chins, and jawlines.

Picasso had been named for the modernistic crosshatch of splotches on her head, the result not of the naturally occurring callosities but of injuries from rope entanglement when she was three years old. The researchers weren't ready to name her daughter yet. She was still so young. The name Bayla would come years later.

But they noted a distinction: unlike the white chins of some right whales, Bayla's was onyx black.

In her first seconds alive, Picasso would have nosed Bayla to the surface for breath—a first tenderness in a year in which she would nurse and cradle and teach Bayla the ways of the sea.

Through the winter researchers saw Bayla and her mother swimming along the coasts of Florida and Georgia, Bayla tucked beside Picasso, safe from great white sharks. The pair began a 1,500-mile trek north as the weather warmed, their boxy bodies and oversized heads moving with unhurried calm, topping out at a poky six miles an hour.

Picasso and Bayla would have crossed shipping lanes off the great ports of the East Coast and swum through the agricultural and industrial runoff of poultry and pork farms in North Carolina and factories in New Jersey and New York. The hazards, documented in accounts including *The Urban Whale,* a book edited by New England Aquarium researchers Scott Kraus and Rosalind Rolland, increased as they moved north. With every mile the risk of collisions with crab and lobster fishing gear would have grown.

Bayla and her mother likely would have stopped in Cape Cod Bay, where their favored food was plentiful in spring, and would have arrived at their destination in Canadian waters in early summer.

The Bay of Fundy, between Nova Scotia and New Brunswick, has among the most dramatic tidal surges in the world. Twice daily, 100 billion tons of seawater rush in and out of the deep rift valley, a result, in native Micmac lore, of a giant whale's tail splash.

Picasso was a regular in the bay, like many right whales, returning in summer to feast on the bay's vast quantities of shrimplike copepods, zooplankton rich in nutrients that scientists suspect are gathered there by the force of the tides.

Bayla's mother would have dived to feast on copepods while Bayla, initially, stayed at the surface.

With a calf's curiosity, Bayla would have taken stock of her new world with right whales' black-and-white vision, registering the life around her in varying shades of gray—the swimming mola molas and basking sharks, the petrels and puffins swooping above in a sky often banked in rolling fog. Upon surfacing, her mother would summon Bayla with ascending moos, known as an upcall, until Bayla returned with a swish of her flippers.

On August 27, biologists noted mud on Bayla's head, a sign of a deep dive—likely a training trip with her mother in the ways of hunting food. Bayla needed to learn well; she was about to enter the most vulnerable period for a right whale. Soon she would be on her own.

After graduation from Cambridge in 1983, Moore moved to Massachusetts, home of Hannah Clark, a forthright music and biology major at Williams College who had been one of his Newfoundland boat mates. The next year they married. Hannah took a job teaching music. Moore enrolled in the MIT and Woods Hole Oceanographic Institution joint PhD program in biological oceanography.

To his colleagues Moore was cordial and persuasive. His craggy face, long and sincere and shadowed by commanding eyebrows, demanded attention when he spoke, which was often in paragraphs. He smoothly politicked through academic logjams, and a reservoir of patience allowed him to maneuver the government bureaucracy that often held the key to funding.

One of his first grants was to study why fish were developing cancer in Boston Harbor. Sewage turned out to be the cause, which led to more research on the pathogenic origin of marine disease. He began showing up at dolphin and whale strandings, looking for samples to collect.

One day when a whale washed up dead on Cape Cod, Moore arrived with his sample kit. He asked the biologist on site why the animal had died.

"You're the veterinarian," said the biologist. "Why don't you tell me?"

Soon after, Moore began regularly arriving home with a rank smell clinging to his clothes. He had become New England's default whale coroner, climbing into carcasses up and down the coastline to determine causes of death. Many of the fatalities were right whales. Some had died from ship strikes or disease, but time and again he found the hulking carcasses tangled in fishing rope. He was bewitched by the right whales. They were megaton creatures who could dive 600 feet, survive on food the size of a grain of rice, and bend their enormous selves to scratch their ears with their flukes—and yet they were regularly succumbing to something so prosaic as fishing rope.

In the fall of 1999, Moore got a call from a NOAA researcher. A team in Lubec, Maine, was trying to cut fishing ropes wrapped around a 10-year-old female right whale. They feared infection had set in where the ropes were cutting her. They wanted to try something new. They wanted him to deliver antibiotics to the whale.

But when Moore arrived in Lubec, he could only watch as a team zigzagged across the water trying to catch up to the distressed right whale. The team tagged her trailing ropes with keggers—buoys meant to slow a whale, similar to the kegs whalers once used to make it easier to close in for the kill. The buoys made the whale thrash harder. The water around her turned frothy white. She was bleeding and vomiting. There would be no getting close enough to deliver antibiotics or disentangle her.

The Coast Guard spotted her a month later off the coast of Cape May. She was hanging below the surface. Once in a while she tried to breathe, until she didn't.

Moore was haunted by the encounter. He couldn't shake the memory of it. The necropsy report turned his stomach: a gill net had sliced a 4.6-foot-wide laceration across her back and carved off a swath of blubber as it sawed toward her tail. The gash exposed both her shoulder blades. Each flipper was incised down to the bone; the left flipper had a 5-inch deep cut and the right flipper had one 7 inches deep. X-rays showed the ropes had deformed her bones and altered the way she swam. When examiners cut the rope, a sharp snap could be heard as the tension finally released from the whale's torso.

Moore felt certain the whale had suffered massive pain. For months, maybe years. The last sighting of the whale before entanglement had been two years earlier, in September 1997. It was beyond what he had imagined. The whale drownings in Bay de

Verde cod nets that he remembered had been comparatively painless—over in a matter of minutes. This was something else. This, he thought, was torment.

He was a marine biologist. Getting exercised about animal pain was dangerous terrain; in the scientific community he could be derided as emotional and unempirical. But he was also a veterinarian. He had taken an oath to prevent and relieve animal suffering. Right whales were venturing into waters humans had claimed for fishing, and they were dying, like roadkill. There had to be a way for humans to coexist with the right whales. Surely he could harness science to find a fix.

Bayla and her mother were seen a final time together shortly before noon on September 21, 2009, in the Bay of Fundy. Bayla had scars on her left flipper and tail, evidence of ropes she somehow had given the slip. Luck had been with her.

She departed the bay sometime in the fall and was seen socializing with bottlenose dolphins and three other young right whales off the coast of Florida in February. It's not clear why Bayla made the trip south, since she was too young to be calving.

Then she did another curious thing. She made no appearance in the Bay of Fundy the next summer. Researchers speculate that she went to the summer home of her grandmother—a maverick who slipped the known migratory routines and social patterns of the right whale community to live by her own code, perhaps summering off Iceland.

There's no real knowing; such is the immensity of the sea and the vastness of what remains unknown about one of the most studied animals in the world.

Figuring out how to stop whales from dying in ropes consumed Moore. Often after dinner with Hannah and the boys, he retreated to his shop, a wood-heated former chicken coop on his island, to theorize.

He wondered if thinner rope might be less injurious, and he directed a student to rig up a machine to test rope widths on blubber. Thinner rope proved more harmful, quicker to cut.

He worked with a Canadian whale biologist to create a model tail, which they used to test a harness they designed. The idea was to wrap the powerful tail of an entangled whale and steady the animal long enough to remove the ropes. In practice, though, the harness didn't latch.

Moore returned to the idea of antibiotics. Perhaps antibiotics could slow infections from lacerations and give an entangled whale a better chance of survival. He'd had great success measuring the blubber thickness of right whales using pole-mounted ultrasonic probes. Perhaps a pole-mounted syringe and needle could work. But needles proved a different matter. If the syringe failed to release from its holder on the pole, a whale researcher holding the pole could end up attached to an angered whale.

At times it felt like science was working against him. Chemists were creating ropes with the strength of steel. Fishermen were opting for the stronger ropes because they lasted longer on the rocky ocean floors.

Regulators were attempting to solve the problem. In some areas of Massachusetts where right whales congregated, they had banned the use of certain gear, including crab and lobster pots.

For whales, the hazard was the rope that runs from a floating buoy to a trap on the ocean floor and the underwater rope connecting traps in a long chain. Regulators at the National Oceanographic and Atmospheric Administration ruled that ropes connecting lobster traps had to sink rather than float, making them less dangerous to swimming whales. Sinking ground rope cost three times as much as the floating rope. It also frayed more rapidly and had to be changed more often, adding up tens of thousands of dollars over a decade for a fisherman.

NOAA also ordered that so-called weak links be used to connect buoys and fishing ropes so that buoys would detach more easily if a whale got entangled. The regulators said the program was a success, pointing out that right whale numbers had increased from 300 to 450 since regulations were phased in starting in the 1990s.

But to Moore it wasn't clear the regulations had any impact on entanglement: right whale deaths from entanglement were on track to double between 2000 and 2014, and the deaths from ropes were getting more gruesome as the ropes got stronger, according to New England Aquarium researchers.

The real answer was off the table. Regulators had decided it wasn't feasible to get rope out of the water column. There was no way they could prohibit lobstering, not in New England. And they had ruled that gear that could free the ocean from ropes—buoys stored on the ocean floor until released by a timer or acoustic signal—was impractical.

Moore believed the regulators could have been bolder had NOAA not also been mandated to consider the economic well-being of industries like fishing that rely on the ocean. Regulators' intentions were good. They committed generous funding to researchers studying entanglement. But Moore felt that regulators ultimately were handicapped by having to serve the conflicting interests of whales and fishermen.

Moore's options were dwindling.

Perhaps the answer lay beyond him, he thought. Maybe it would take the market. Whole Foods could take a cue from Massachusetts. Lobsters caught in state waters were sold with whale-logo-stamped green bands on their claws to show that fishermen had used sinking lines to connect their lobster traps. If fishermen adopted additional and more effective whale-safe techniques, chains like Whole Foods could market lobsters as whale-safe, a kind of fair-trade movement for whales.

But Moore didn't have a clue about where to begin with that. He was a scientist, not a consumer advocate. He had to focus on what he knew.

He had one more idea.

One day in the middle of winter 2006, Moore picked up his phone and dialed a number halfway around the world.

On Christmas Day in 2010, a Florida Fish and Wildlife aerial team was making a regular survey of right whales. Off the coast of Jacksonville they spotted a right whale. Visibility was low. But they could see ropes wrapped tightly around it. Rescue teams deployed. Conditions were so bad rescuers could only attach a tracking device to the end of rope trailing the whale so her location could be followed.

Four days later, when the weather cleared, rescuers found the whale 30 miles south in the St. Augustine Inlet. They tried but couldn't get good cuts on the rope. The next day they sheared a large loop of rope. But the whale was panicked and evasive and wouldn't let them get close enough to make the critical cuts to the rope in the whale's mouth that held the complex weave in place.

New Year's Day passed. The whale swam north to Fernandina Beach. Rescuers noted the whale was dangerously thin.

Word came back from the New England Aquarium. The whale's markings matched those of a whale in the right whale catalog.

It was Bayla.

Moore liked the voice he heard on the other end of the phone line that day in 2006. Trevor Austin had a matter-of-fact Kiwi delivery. He was an engineer whose company in New Zealand made equipment for tranquilizing animals.

Moore said he had an idea. He wanted to sedate a free-swimming whale so that rescuers could get close enough, for long enough, to remove ropes entangling it. He needed a ballistic system that would send 60 milliliters of sedative flying through the air with enough force to penetrate a whale's fibrous blubber with a 12-inch needle.

Austin was silent. The largest-capacity animal syringe held a tenth of what Moore wanted.

"We'll give it a go," Austin said.

Moore sent him $25,000 that NOAA had supplied for the project. A year later, on a raw March day, Austin arrived at Logan Airport carrying a black Pelican case packed with anodized aluminum tube syringes, stainless steel needles, and a dart rifle.

Moore and Austin drove to a range on the Cape. Moore had a cooler with three squares of dolphin blubber. Tacked together, they measured the same depth as a right whale's. Moore perched the blubber against a hay bale.

Austin took a shot. A blank .22 cartridge sent the dart exploding out of the chamber. The dart bounced off the blubber.

Like a bulletproof vest, Austin thought.

Austin fired again. This time the dart entered the blubber but promptly bent in half. Austin and Moore huddled. The syringe's momentum had continued after the needle entered the blubber, taking the needle along and bending it. The needle needed more resilience. They retreated to Moore's lab at Woods Hole and glued a carbon fiber tube to the needle's stainless steel husk.

The next day they drove back to the range. Austin took a shot. A perfect strike.

Moore got a live test case six months later. A mother and calf humpback had gone astray on their migration north and swum 90 miles inland in the Sacramento River Delta, through three bays and past five bridges. They had wounds, likely from a ship's propeller, and the fresh water was degrading their skin. They needed antibiotics. Moore was called out to California. He loaded the medications into the dart rifle's syringe. A colleague aimed and fired. Within a week the whales, successfully treated, had regained enough strength to make their way back to the open ocean.

Moore had his device.

On a January day in 2011, Moore was sitting in his office at Woods Hole when the e-mail arrived. Was he available to sedate the calf of Picasso? Yes, he replied. He was on his way.

In a hotel room in New Smyrna, Moore and the fleet of biologists from across Florida, Georgia, and Massachusetts reviewed the details of the operation. The meeting went long and sleep was short, but the next morning Moore's mind whirred with possibility as the radio of the overhead airplane reported that Bayla was surfacing.

Bayla's back glistened as it moved across the waterline on the January morning, a black sheath divided by the ropes leading out of her mouth.

The man standing next to Moore on the tower of the Zodiac cocked the dart rifle tipped with the sedative-filled syringe. "We're live," Jamison Smith said. Moore was a good shot, but Smith, who helped direct the federal effort to stop whale entanglements, was better. He'd grown up hunting ducks and other waterfowl with four brothers in Florida.

Don't miss, Moore thought as he gave the command to shoot.

The report was a sharp crack. A splash erupted at the waterline next to Bayla's torso. An orange buoy tied to the syringe for tracking fell to the water. The buoy jerked, then began moving. The dart was traveling with Bayla.

"The f–ing thing worked," Moore said, his voice rising and surrendering to surprised wonder.

The radio erupted with excited chatter from the other boats and the overhead plane. As they tracked Bayla diving and surfacing, then diving and surfacing again, elation bubbled. The sedative hadn't killed her. But there was work yet to do. "Back away," Moore called to the driver of his boat.

The clock was ticking. The sedation would last 90 minutes.

Soon a more nimble inflatable craft moved in.

The boat was piloted by Chris Slay, a Georgia biologist and motorcycle racer. Mark Dodd, a lifelong surfer and another Georgia biologist, held a carbon-fiber pole fitted with a knife that Slay had designed for cutting tight entanglements like Bayla's. Instead of a traditional hook that slid under the rope to make a cut, a spring-loaded mechanism sent a blade plunging at the rope from above. Dodd was on his knees, scooted into the bow, like the nose gunner of a B-17.

Dodd and Slay were old hands at freeing whales from ropes. They had pursued entangled whales up and down the Georgia and Florida coasts, at times only to have the whales disappear before they could get a single cut.

Slay motored behind Bayla, guided by two beach-ball-sized orange buoys that had been attached to her trailing fishing ropes that morning. Her pace and cadence had to be understood before the chase could begin. What Slay saw astonished him. Bayla was swimming in a straight line. None of the sharp turns he was used to with right whales dodging rescuers. Every five minutes she came up to breathe. No deep, unpredictable dives.

She was a perfect target. She was theirs to lose.

Slay gave Dodd the signal. As Bayla surfaced, Slay gunned the motor, closing the distance between Bayla and the boat. Dodd punched the ropes with the knife, but the ropes didn't give. Slay backed the boat off. They tried again. Another miss.

Something was misaligned, Dodd shouted over the whine of the engine. They couldn't afford another flawed rally. "I'm not sure anything was cut. Honestly, I'm not sure."

"Focus as much as you can on exactly where you want to hit it," Slay said.

The aerial team radioed. Bayla was rising. Dodd leaned over the gunwale. He could see a shadow a few meters off the right bow. "Her bonnet's right there! She's coming up. See it right there, Chris?"

Slay opened the throttle. The boat clipped right. Dodd hoisted the knife. Bayla's head emerged and water cascaded down her sides. A spangled spray of phlegmy blow blasted Dodd's eyes and nose.

Fundamentals, Dodd thought.

He pictured the matrix of ropes, their loops and twists. As the spray cleared, the ropes appeared and were level with his face. He heaved forward with a grunt and thrust the pole.

If Bayla felt anything, she gave no indication. But before the force of the strike repelled him to the floor of the boat, Dodd glimpsed the ropes slacken.

Slay whipped his head around. Two orange balls bobbed in place behind the boat.

"The buoys are dead in the water," Slay said.

The ropes had fallen free.

As the sun descended, the inflatables steered into harbor. Fatigue was settling. There would be time later for the team to deconstruct the day's success, but Moore needed to know one thing. He approached Slay.

"Did the sedation make a difference?" Moore asked.

Slay smiled. "Hell yes."

For six days Moore's e-mail pinged with daily updates on Bayla's coordinates as she swam south down the Florida coast. The information came from a temporary satellite tag that the disentangling team had attached.

Then, as planned, the tag had fallen away and Moore's e-mail had gone dark with news of Bayla.

Her whereabouts were unknown.

On February 1, Moore leaned against the office doorway of his graduate student, Julie van der Hoop. "There's a dead whale in Florida," he said.

She had followed the news of Bayla like everyone in the right whale community, asking Moore each morning if another ping of her coordinates had come in. Her face darkened.

"Is it the calf?" van der Hoop asked.

"I don't know," Moore said. "There's a necropsy scheduled. Do you want to come?"

Moore and van der Hoop arrived in St. Augustine the next day. The following morning they watched as the sun rose over a young right whale beached on the sand.

She had a black chin.

A team of researchers cut away Bayla's shark-mauled blubber with long knives and examined her internal organs. In her mouth they discovered rope that Dodd and Slay hadn't gotten. It was so deeply embedded, new tissue had grown over it, "like a pig in a blanket," van der Hoop would later observe in her journal.

By day's end the team determined that Bayla had died from

severe emaciation and lacerations caused by hundreds of feet of 7⁄16-inch-diameter floating polypropylene rope that connected traps or pots—the sort that NOAA had attempted to restrict.

Beyond that, there wasn't any more to be known. The rope could have come from off the coast of New England, or perhaps Canada; by the time Moore's team had cut it away, she was too weakened to survive.

There would be no decorous burial for Bayla; her size defied it. An excavator scooped her muscle and soft tissues into a hole dug in the mucky sand. The loader then piled Bayla's bones onto a truck destined for Atlanta, where her skeleton would be reassembled for display at the Georgia Aquarium and she would be given the Hebrew name Bayla, meaning "beautiful."

When the work was done, Moore held a needle. It was the needle that researchers had fired from the rifle to deliver antibiotics to Bayla after her sedation. It was bent at an 80-degree angle. He suspected it had caused Bayla more pain. There were lessons to be learned from why it bent. He would write a paper. His peers would review it. A journal would publish it.

But that was for another day.

For now, Moore cried.

The needle sat on his desk for a year taunting him.

The paper was hard to write. Harder than any other. When it was done, there was relief. But the relief soon was replaced by creeping doubt.

A colleague e-mailed him thanking him for his efforts to save the right whales from entanglement. Moore replied that he wasn't sure the thanks was due. He still had no solution. After all these years, he still didn't get it.

A bleak realization had settled, he wrote. "I've failed."

The winds in the Bay of Fundy were steady and Moore hoisted the mainsail of the *Rosita,* the sailboat he and Hannah named for a whaling station in the South Atlantic that had been planned but never built. Moore liked to think of the *Rosita* as embodying the spirit of whales spared the harpoon.

Every few minutes Moore whipped his head right or left, drawn by the chuff of spouting water. Grand Manan Island spread across the western horizon. Ahead, right whales lolled at the waterline, breathing hard after what must have been deep dives for food. There were dozens of right whales in the bay.

Almost to a one, they had fishing rope scars.

Moore and Hannah were at the end of a summer vacation. It was August 2014, more than three years since Bayla's death. He had continued whale research but often felt he was going through the motions. He was due back in the office at Woods Hole in a few days, but not certain of what he was returning to do.

Over vacation, an idea had begun swatting at him, one his younger self would have considered heresy. Science had been superb at documenting the problem of entanglement. But science had not been good at finding a solution to end it.

How many papers had he written? How many necropsies had he performed? How many ideas had led nowhere?

Sedation had proved workable, but inevitably came too late for whales like Bayla. Moore and others had concluded that prevention was the only answer in cases like hers.

There were new regulations coming online. Fishermen soon would have to attach a minimum number of traps to a buoy line and they would have to better mark their ropes. And the areas in Massachusetts where right whales congregated were to be closed to gear such as traps and pots for longer periods.

But fishermen were protesting, and NOAA was revisiting some of the new rules.

Moore always thought that if dogs walked around the city of Boston with bleeding lacerations, people would become outraged and demand that the source of injury be stopped.

Whales swam unseen with their wounds.

He was 57. Retirement was approaching. But there was time yet.

Maybe if he could communicate what he had felt all those years ago. If people could feel what he felt when he heard the whales singing in his dreams, maybe then they would come to share his heartache, and wake to the need to do more.

Mist spouted in front of the *Rosita.* Moore climbed onto the prow. Sun was splintering through the clouds, and the slanted rays met the water in bangles of light. Somewhere out there Picasso swam. Aquarium researchers had spotted her in the bay. Perhaps she was the right whale in front of him, dunking its head and driving its tail into the air until it was perpendicular with the surface, like a salute to the terrestrial world.

"There's something about a right whale's tail that's just gorgeous," Moore mused. "Michelangelo could have sculpted it."

Moore rested his body against the mast. The ocean spanned before him.

"We're surrounded by right whales," he said.

MICHAEL SPECTER

Partial Recall

FROM *The New Yorker*

ONE MORNING EVERY spring, for exactly two minutes, Israel comes to a stop. Pedestrians stand in place, drivers pull over to the side of the road, and nobody speaks, sings, eats, or drinks as the nation pays respect to the victims of the Nazi genocide. From the Mediterranean to the Dead Sea, the only sounds one hears are sirens. "To ignore those sirens is a complete violation of the norms of our country," Daniela Schiller told me recently. Schiller, who directs the laboratory of affective neuroscience at the Mount Sinai School of Medicine, has lived in New York for nine years, but she was brought up in Rishon LeZion, a few miles south of Tel Aviv. "My father doesn't care about the sirens," she says. "The day doesn't exist for him. He moves about as if he hears nothing."

Sigmund Schiller's disregard for Holocaust Remembrance Day is perhaps understandable; he spent the first two years of the Second World War in the Horodenka ghetto (at the time in Poland, but now in Ukraine) and the next two hiding in bunkers scattered across the forests of Galicia. In 1942, at the age of 15, he was captured by the Germans and sent to a labor camp near Tluste, where he managed to survive the war. Trauma victims frequently attempt to cordon off their most painful memories. But Sigmund Schiller never seemed to speak about his time in the camp, not even to his wife.

"In sixth grade our teacher asked us to interview someone who survived the Holocaust," Daniela Schiller said. "So I went home after school. My father was at the kitchen table reading a newspaper, and I asked him to tell me about his memories. He said nothing.

I have done this many times since. Always nothing." A wan smile crossed her face. We were sitting in her office, not far from the laboratory she runs at Mount Sinai, on Manhattan's Upper East Side. It was an exceptionally bright winter morning, and the sun streaming through the window made her hard to see even from a few feet away. "I long ago concluded that his silence would last forever," she said. "I grew up wondering which of all the horrifying things we learned about at school the Germans did to him."

Slowly, over the years, that silence closed in on her. "It wasn't so much a conscious thing," she said. "But I grew up with that fear in the background. What was he hiding? Why? How do people even do that?" The last question has, to a large degree, become the focus of her career: Schiller studies the intricate biology of how emotional memories are formed in the brain. Now 41 and an assistant professor of neuroscience and psychiatry at Mount Sinai, she specializes in the connection between memory and fear. "We need fear memories to survive," she said. "How else would you know not to touch that burner again? But fear takes over the lives of so many people. And there is not enough that we can do about it."

More than 5 percent of Americans have experienced some form of post-traumatic stress disorder; for combat veterans, like those returning from Afghanistan and Iraq, the figure is even higher. Millions of others suffer from profound anxiety, debilitating phobias, and the cravings of addiction; those emotions appear to be formed in the same neural pathways, which means that a successful treatment for one condition might also work for others. Behavioral therapies, even those which work initially, often fail. Relapses are common, and the need for more successful treatments has never been so acute. New approaches are hard to develop, though, because most of what is known about the human brain has come from studying the neurons of other animals. One can't simply stick a needle into somebody's brain, grab a few neurons, drop them in a nutrient bath, and see what happens. PET scans and functional magnetic resonance imaging machines have helped address the problem; they permit neuroscientists to monitor metabolic changes and blood flow in the human brain. But neither of them can measure the activity of neurons directly.

Even so, Schiller entered her field at a fortunate moment. After decades of struggle, scientists had begun to tease out the complex molecular interactions that permit us to form, store, and recall many different types of memories. In 2004, the year Schiller received her doctorate in cognitive neuroscience, from Tel Aviv University, she was awarded a Fulbright fellowship and joined the laboratory of Elizabeth Phelps at New York University. Phelps and her colleague Joseph LeDoux are among the nation's leading investigators of the neural systems involved in learning, emotion, and memory. By coincidence, that was also the year that the film *Eternal Sunshine of the Spotless Mind* was released; it explores what happens when two people choose to have all their memories of each other erased. In real life it's not possible to pluck a single recollection from our brains without destroying others, and Schiller has no desire to do that. She and a growing number of her colleagues have a more ambitious goal: to find a way to rewrite our darkest memories.

"I want to disentangle painful emotion from the memory it is associated with," she said. "Then somebody could recall a terrible trauma, like those my father obviously endured, without the terror that makes it so disabling. You would still have the memory, but not the overwhelming fear attached to it. That would be far more exciting than anything that happens in a movie." Before coming to New York, Schiller had heard—incorrectly, as it turned out—that the idea for *Eternal Sunshine* originated in LeDoux's lab. It seemed like science fiction, and for the most part it was. As many neuroscientists were aware, though, the plot also contained more than a hint of truth.

Concepts of memory tend to reflect the technology of the times. Plato and Aristotle saw memories as thoughts inscribed on wax tablets that could be erased easily and used again. These days we tend to think of memory as a camera or a video recorder, filming, storing, and recycling the vast troves of data we accumulate throughout our lives. In practice, though, every memory we retain depends upon a chain of chemical interactions that connect millions of neurons to one another. Those neurons never touch; instead, they communicate through tiny gaps, or synapses, that surround each of them. Every neuron has branching filaments, called dendrites, that receive chemical signals from other nerve cells and send the information across the synapse to the body of the next cell. The typical human brain has trillions of these connections. When we learn something, chemicals in the brain strengthen the synapses that connect neurons. Long-term memories, built from new proteins, change those synaptic networks constantly; inevitably, some grow weaker and others, as they absorb new information, grow more powerful.

Memories come in many forms. Implicit, procedural memories —how we ride a bike, tie our shoes, make an omelet—are distributed throughout the brain. Emotional memories, like fear and love, are stored in the amygdala, an almond-shaped set of neurons situated deep in the temporal lobe, behind the eyes. Conscious, visual memories—the date of a doctor's appointment, the names of the presidents—reside in the hippocampus, which also processes information about context. It takes effort to bring those memories to the surface of awareness. Each of us has memories that we wish we could erase and memories that we cannot summon no matter how hard we try. At NYU and other institutions, scientists have begun to identify genes that appear to make proteins that enhance memory and genes that clearly interfere with it. Both kinds of discovery raise the tantalizing, if preliminary, hope of a new generation of drugs, some of which could help people remember and some that might help them forget.

Until memories are fixed, they are fragile and easily destroyed. Who has not been interrupted while trying to remember a phone number or an address? That memory almost invariably slips away, because it never had time to form. (This also explains why accident victims often have trouble recalling events that occurred just before a car crash or other severe trauma.) It takes a few hours for new experiences to complete the biochemical and electrical process that transforms them from short-term to long-term memories. Over time they become stronger and less vulnerable to interference, and, as scientists have argued for nearly a century, they eventually become imprinted onto the circuitry of our brains. That process is referred to as consolidation. Until recently few researchers challenged the paradigm; the only significant question about consolidation seemed to be how long it took for the cement to dry.

For years, though, there have been indications that the process is less straightforward than it seems. In 1968 a team at Rutgers, led by Donald J. Lewis, published the results of an experiment in which rats were conditioned to retrieve memories that had presumably been stored permanently. First the scientists trained the rats to fear a sound. The next day Lewis played the sound again and followed it immediately with a shock to the head. To his surprise, the rats seemed to have forgotten the negative association; they no longer feared the sound. That seemed odd; if the memory had truly been wired into the rat's brain, a mild shock shouldn't have been able to dislodge it. The experiment wasn't easily repeated by others, though, and few neuroscientists paid much attention to such a singular and contradictory finding.

Not long afterward, in seemingly unrelated research, the psychologist Elizabeth Loftus, now at the University of California at Irvine, embarked upon what has turned into a decades-long examination of the ways in which misleading information can insinuate itself into one's memory. In her most famous study, she gave two dozen subjects a journal filled with details of three events from their childhoods. To make memories as accurate and compelling as possible, Loftus enlisted family members to assemble the information. She then added a fourth, completely fictitious experience that described how at the age of five each child had been lost in a mall and finally rescued by an elderly stranger. Loftus seeded the false memories with plausible information, such as the name of the mall each subject would have visited. When she interviewed the subjects later, a quarter of them recalled having been lost in the mall, and some did so in remarkable detail.

"I was crying and I remember that day . . . I thought I'd never see my family again," one participant said in a taped interview (available on YouTube). "An older man approached me . . . He had a flannel shirt on . . . I remember my mom told me never to do that again." These assertions were delivered with a precision and a certainty that few people could have doubted, except that there was no man in a flannel shirt and no admonition from the subject's mother. Memory "works a little bit more like a Wikipedia page," Loftus said in a recent speech. "You can go in there and change it, but so can other people."

Loftus has been vilified for demonstrating that even the most vivid and detailed eyewitness accounts—a "recovered memory" of sexual abuse, for example—can be inaccurate or completely false. "She changed the world," Elizabeth Phelps told me recently when we met in her office at NYU, where she is the Silver Professor of Psychology and Neural Science. "The notion of the unreliability of memory has changed courtrooms in America, and it is completely owing to Elizabeth's persistence in the face of a very harsh backlash."

Loftus's results raised a fundamental question about the biology of the brain: if misinformation can be incorporated so seamlessly into a person's recollection of an event, what becomes of the original memory? Is it completely overwritten, or merely adjusted somehow, layered with a new trace?

In the decade following Loftus's experiment, an answer began to emerge, as LeDoux, Phelps, and others slowly mapped the neural circuitry responsible for many types of memory, particularly memories associated with fear. They began to entertain the idea that in order for an old memory to be recalled, it had to retrace the pathways in which it originated, and that under certain circumstances the memory seems to change. Scientists called that reconsolidation. But reconsolidation, with its eerie implication that our memories are inauthentic or transitory, was highly disputed. To many scientists, while the idea was fascinating, it remained farfetched.

By 1996, LeDoux's lab had demonstrated that fearful memories were particularly durable, but also that when certain parts of the amygdala were destroyed, those fears disappeared. That year Karim Nader joined the laboratory as a postdoctoral researcher. Not long afterward he attended a lecture given by Eric Kandel, the Columbia University neuroscientist who in 2000 received a Nobel Prize for his research into the physiological basis of memory. Kandel spent decades demonstrating how neurochemicals form short-term memories and how more permanent memories are then consolidated into various parts of the brain. Without his findings, none of the research into emotional memory that followed would have been possible.

In the early 1960s, Kandel decided to conduct classic Pavlovian conditioning studies on aplysia, or sea slugs, which have relatively few neurons. More important, aplysia possess what Kandel has described as the "largest nerve cells in the animal kingdom. You can see them with your naked eye." That made them easy to manipulate in a laboratory. Kandel removed neurons and placed them in a petri dish. By stimulating the neurons with an electrode, he was able to map the entire neural circuit required to cause a common reflex. (The reflex he chose forces the slug's gills to retract when they are disturbed, in much the same way that a threatened porcupine will raise its quills.)

Scientists were already aware that making a memory requires chemical activity in the brain. But neurons are programmed by our DNA, and they rarely change. On the other hand, synapses, the small gaps between neurons, turn out to be highly mutable. Synaptic networks grow as we learn, often sprouting entirely new branches, based on the way that chemical messengers called neurotransmitters pass between neurons. "The growth and maintenance of new synaptic terminals makes memory persist," Kandel wrote in his book *In Search of Memory: The Emergence of a New Science of Mind* (2006). "Thus, if you remember anything of this book, it will be because your brain is slightly different after you have finished reading it."

Nader was thrilled by the idea that one could watch an organism form a memory. "I was not trained as a neuroscientist in memory or in consolidation," he told me recently on the phone from McGill University, where he is now a professor of psychology. "Kandel talked about the physiology of the neuron on the most basic level, and I was amazed. But I didn't understand why a thing like that—the complete chemical production required to form a memory—would happen just once. I looked at the data and thought, What makes us so certain that after our memories are formed, they are fixed forever?"

The prospect that a memory might be altered simply by being recalled was heretical; LeDoux urged Nader not to waste his time. But he was determined, and LeDoux didn't interfere. Early in 1999, Nader and his colleagues devised an experiment in which they trained a group of rats to fear a tone. Conditioning, for rats and most species, including ours, is relatively straightforward: a researcher will pair a neutral stimulus like a tone or a color with something unpleasant, usually a shock. The results are quick and definitive; replay the tone, even without the shock, and the rat will freeze in place, crouching as low as it can. Its fur will stand on end, and its blood pressure will soar. The next time the rat (or human) hears the tone, the electrical circuitry in its brain responds as powerfully as if it were also experiencing the shock, and the synapses associated with that memory will grow stronger.

After teaching the rats to fear the tone, Nader waited 24 hours, to give their memories time to consolidate. Then he played the tone again and injected the antibiotic anisomycin into the rats' lateral amygdala, the area that houses fearful emotions. Anisomycin has been shown to prevent neurons from producing the proteins necessary to store a memory. If memories are formed just once, Nader reasoned, the drug should have no effect. "The idea," he said, "was that if a new set of proteins was required, then the drug should prevent the memory from being recalled." That is exactly what happened. Rats that received the drug within four hours of recalling the memory forgot their fear. Two weeks later, when Nader again tested the rats, those with blocked memories responded as if they had never heard the tone. Rats in two control groups—one of which received no shot, the other of which received a placebo injection that did nothing to prevent synapses from making new proteins—remained terrified.

Nader's data could not have been clearer, or more unsettling. He had demonstrated that the very act of remembering something makes it vulnerable to change. Like a text recalled from a computer's hard drive, each memory was subject to editing. First you have to search the computer for the text and then bring it to the screen, at which point you can alter and save it. Whether the changes are slight or extensive, the new document is never quite the same as the original.

Many people in the field treated Nader's findings with contempt. James L. McGaugh, of the University of California at Irvine's Center for the Neurobiology of Learning and Memory, and one of the nation's leading neuroscientists, argued, like most of his colleagues, that once long-term memories are established, they are there to stay. "Occasionally the seduction of simplicity embarrasses the field," McGaugh and two colleagues wrote at the time. He compared work on reconsolidation like Nader's to notoriously inaccurate research, begun in the 1960s but long since debunked, suggesting that it was possible to transfer intelligence from one animal to another through "memory molecules." "We should be careful not to laugh in retrospect at such ideas," McGaugh wrote, "if we remain attracted to other more contemporary simple explanations of the complex phenomena of learning and memory."

Scientists around the world soon set out to repeat Nader's study, and the results of experiments in dozens of species, from fruit flies to mice, supported his conclusions. The dogma of consolidation made no sense. It is one thing, of course, to erase a fear created in a laboratory and applied to rats, and another to do it with humans. Daniela Schiller was in Israel at the time, finishing her doctorate. Using an animal model, she had studied the relationship between emotion and neural circuitry in schizophrenia. When Schiller learned of Nader's findings, she wondered if it would be possible to reactivate a traumatic memory in humans and then block the fears associated with it, much as Nader had done in rats. With her father's advancing age never far from her mind, she became determined to find out.

Daniela Schiller is tall and trim, with steel-blue eyes and dark-blond hair. When she strides through her laboratory at Mount Sinai, Schiller—nearly always dressed in understated outfits designed by her sister, Yael, in Tel Aviv—carries herself more like a Middle European aristocrat than like a woman who grew up in a scruffy suburb of Tel Aviv. Schiller's mother is Moroccan, and she says that her father, who suffers from emphysema, sounds like a sort of Polish Darth Vader. "You hear him before you see him," she said. Schiller is the youngest of four children; her two older brothers and her sister stayed in Israel, and her parents still live in the house where she grew up. Science always appealed to Schiller. "I would mix sand from the backyard with all sorts of materials I found at home and turn it into weird solids and liquids," she told me. The concoction "looked like a top-secret chemistry set, in my little mind, so I asked a neighbor to hide it in her backyard. After a few days she asked me to take it back. She was worried it might blow up or something."

The winter Schiller started working at NYU, she noticed her boss, Joseph LeDoux, playing guitar at a Christmas party with Tyler Volk, a professor of biology. Schiller is a drummer, and she soon found a lab mate who played bass. The four formed the Amygdaloids, which, despite the gimmicky name, is far better than one might suspect of a band born in a brain lab. At NYU, Elizabeth Phelps asked Schiller to work on a study that might determine whether humans would respond the way rats did to Nader's experiments. But the drug used for rats was far too toxic to use on people. Instead Schiller used propranolol, a common beta blocker that, because it latches on to receptors in a variety of proteins, has been shown to interfere with the formation of memories. She applied to the university for permission to carry out the experiment and waited for a response; she has not yet received one.

During a laboratory meeting, however, Schiller's colleague Marie Monfils mentioned that after behavioral training, a group of rats in one of her experiments seemed to lose their fear. The finding was serendipitous; Monfils had originally been studying something else. But the comment provided Schiller with what she describes as her "eureka moment." Until then memory reconsolidation had been blocked only by physical intervention, either drugs or electric shocks. If, as scientists have suggested, reconsolidation evolved so that memory could be augmented with new information, then behavior modification ought to have the same effect as a drug. "I suddenly realized that we had never tested that theory," Schiller told me. Monfils agreed to carry out a behavioral study of rats, and Schiller would do the same with humans.

The theory was borne out by both experiments. Schiller trained 65 people to fear a colored square by associating it with a shock. The next day, the sight of the square alone was enough to revive their fearful reactions. Then Schiller divided the subjects into three groups. By presenting the squares many more times with no shock, she attempted to teach them to overcome their fear. That is called extinction training. The results were dramatic: people who saw the squares within 10 minutes of having their memories revived forgot their fear completely. The others, who were not shown the squares again until hours later, remained frightened.

Schiller's study, which was published in *Nature* in 2010, offered the first clear suggestion that it might be possible to provide long-term treatment for people who suffer from PTSD and other anxiety disorders without drugs. And the effect seemed to last; a year later, when the researchers tested the subjects again, the fear response still had not returned.

Schiller moved to Mount Sinai in 2010. Since then she has pursued three central goals in her research: tracing the neural mechanism, or signature, that causes memory to update in the human brain; determining whether drugs might work safely in humans; and establishing a protocol that therapists could use to treat patients. (Scientists have already found that behavioral interference during reconsolidation appears to alter glutamate receptors in the amygdala, which might explain how memories are rewritten during the treatment.)

On a particularly harsh winter morning in February, I joined Schiller and one of her postdocs, Dorothee Bentz, at the Mount Sinai School of Medicine's Brain Imaging Core. Despite its impressive, *Matrix*-like name, the Core is a closet-sized room filled with computers and electrical machinery. The gauges and ominous-looking dials seem to belong on an old radio set. Bentz attached electrodes and sensors to my arms and to my right wrist, told me to take a deep breath, and then started ramping up the voltage. I watched the meter as the needle jumped.

"Do you feel that?" Schiller asked, somewhat remotely. "It's twenty volts, a small charge." I said no. She moved the lever to 30. Yes, but only barely, I told her. Finally, at 40 volts, I began to feel the shock. It was by no means a dangerous level; nonetheless, it was a sensation that few people would welcome. Schiller was planning to do to me what she had spent so much time doing to others: teach me to fear a meaningless symbol. Colored spheres began to float onto a computer screen in front of me, in no particularly discernible pattern: just a random, rapid-fire procession—purple, yellow, and blue. It didn't take long to realize that nearly every time a blue sphere appeared a shock would follow; by the time I felt the voltage, my pulse and heart rate had already spiked in anticipation. The shock itself quickly became superfluous.

The day after learning to fear the spheres, Schiller's subjects see them again many times—but without the accompanying shock. "If you present a negative memory over and over again, without anything bad happening, it is possible for most people to overcome the fear," Schiller explained. Extinction training has for a long time been one of the principal treatments for many phobias and fears; psychiatrists refer to it as exposure therapy. The more you see something, the less it scares you, and the less it scares you, the more able you are to deal with it. There has always been a problem, though, in using extinction to treat people who have experienced profound trauma: the process leaves them with a pair of memories: blue sphere predicts shock; blue sphere doesn't predict shock. Over time the two memories can compete for expression. That is a significant characteristic of anxiety disorder. People will be fine for months or years, but if they encounter a particularly stressful situation, the fear memory often overwhelms the calm memory.

Schiller's study demonstrated that the competing memories can become one. "If we zap it at just the right time, there are no new memories," she told me with a look of restrained satisfaction. "There is a different memory. You will still know what happened, and the information will be available to you. But the emotion will be gone."

Schiller has applied for funding to continue the research. Deep budget cuts have made it harder to get money than ever before, though, and her initial three-year grant at Mount Sinai has nearly come to an end. I asked what would happen if she received no money. "I'm back on the street," she said, shrugging. "But I believe we can find a way to make PTSD less terrible. From the research perspective, you really do get very optimistic. Of course, I am careful not to try and overhype it. Translating research into better human lives is never easy."

Not long after my fear test, I took the train to Philadelphia to speak with Edna Foa, who is the director of the Center for the Treatment and Study of Anxiety, at the University of Pennsylvania Medical School. Foa is one of the nation's leading experts on the psychopathology of anxiety disorders, and she has written widely on PTSD. We met in her office at the medical school, which looks onto the oddly serene urban landscape of Center City. I asked if she thought scientists would ever really be able to write the pain out of a patient's mind.

"That is the critical question," she replied, stressing that she is a clinician, not a neuroscientist. "This is the most exciting prospect I think I have ever seen for treating people with severe anxiety-based disease. It isn't easy to banish demons caused by war, trauma, and rape." Freud argued that repressed memories, blocked unconsciously, were like infections, capable of deepening and festering unless they were brought to the conscious mind and resolved. Many psychiatrists have taken the opposite approach. "There has always been a group that says we could reignite a trauma by asking people to deal with the memory," Foa said. "In this thinking, keeping the memory suppressed was actually better. That was a strong belief in the early era of psychiatry: Put it behind you. Don't deal with it. Go on with your life. The idea behind counseling was to soothe the patient, to find ways to make him as comfortable as possible."

Only in the past decade have researchers determined that while the original memory may be inhibited, it doesn't vanish. Foa said that the idea of rewriting memories, rather than destroying them, appealed to her. But she added that reconsolidation raises a paradox: in order to update our most painful memories, we have to revisit them. That is never easy to do. Foa described a patient who was raped more than a decade ago by her boyfriend and several of his friends. She suffered badly from PTSD, found it impossible to maintain relationships, and had recently entered therapy. "Instead of asking herself what actually happened, she would immediately say it was all her fault," Foa said. "She always said the same thing: 'I didn't fight them. If I had, they would have stopped.'

"But she never dealt with it, and that is why she had PTSD," Foa went on. "We asked her to tell the story of that New Year's Eve and repeat it many times." As people work through the story again and again, they learn to distinguish between remembering what happened in the past and actually being back there. For people with PTSD, this distinction is not easy to make. The next step was to bring those memories to the surface—and when, finally, the woman did that, she realized that her terror and her rape were not her fault.

I asked Foa if she had considered the ethical complexity involved in tampering with a person's memory. "Of course," she replied. "But you do have to look at the whole picture. We are talking about helping people who have been severely traumatized, and in many cases they are unable to function. Nobody is suggesting that we rewrite the memory of someone who had a bad date or a fight with his mother."

In practice it may be hard to draw an ethical line that would satisfy patients, doctors, and the public. Few people would deny effective treatment to victims of severe brutality. But any treatment available to those who need it will almost certainly be available to others. "Memory erasure remains a possible but unproven hypothesis," Joseph LeDoux has written, adding that editing memories "is definitely possible and has broad implications. We are nothing without our memories, but sometimes they also make us less than we could be . . . Although some ethicists argue that memory should not be tampered with, every special date and anniversary, every advertisement, every therapy session, every day in school is an effort to create or modify memory. Tampering with memory is a part of daily life. If we take a more realistic view of just how much we mess with memory, the dampening of memories that produce emotional responses in traumatized individuals might seem less malevolent."

Reconsolidation has already been shown, in promising if limited research, to help treat drug addiction. Addicts are compelled by the same persistent emotional memories that drive other disorders. "The biggest problem for most addicts is how to deal with relapse," Schiller told me. "Let's say somebody is drug-free and then goes and hangs out with friends at a park. He might see a cue associated with his drug use, and that will induce a craving that will cause him to seek the drug." Reconsolidation presents a chance to disrupt that process; you don't lose the memory—you just lose the pleasant feeling it creates.

The idea is simple enough: you cannot be addicted to a desire that you don't remember. Jonathan Lee, a behavioral neuroscientist now at the University of Birmingham, in England, has already put that notion to a test. He used Pavlovian conditioning to induce cravings in rats, by pairing light with a narcotic. The next time he showed the animals the light, they automatically reached for the drug. But, as was the case with Nader's experiments, when Lee interrupted the process of reconsolidation, the association disappeared. Researchers in the U.S. and China have had similar success with human addicts. Once again, timing was critical: the effect worked only if extinction training took place within 10 minutes of retrieving the old memory. "If you block that association, you can erase the craving," Schiller said. "This is the first time we have seen a treatment like that lead to a cessation of addiction." Even six months later, the addicts showed no sign of relapse, suggesting, as with Schiller's work, that when fearful memories are disturbed at the right moment, the fear may be gone for good.

At the age of 88, Sigmund Schiller, with a mustache, goatee, and nearly bald pate, looks like an aging Lenin. These days he spends most of his time tending the small, immaculate garden behind the ranch house that he and his wife, Yaffa, have lived in for nearly half a century.

I had come to his house, in this sunny spot between Ben Gurion

Airport and the Mediterranean coast, for an unlikely reason: not long ago, after decades of unwavering silence, Schiller spoke about his Holocaust experience. It happened once, and he says that it won't happen again. But his words were filmed for a documentary. I had watched it with Daniela Schiller in Brooklyn, at the home of the director, Liron Unreich; the day I visited Rishon LeZion, Schiller's parents were about to see the film for the first time. Unreich is a multimedia artist and a cofounder of the Ripple Project, which explores the multigenerational effects of the Holocaust through short documentaries. Schiller, who took part in the film, had been astonished when her father began to talk to Unreich and his crew. "To say I never expected it would be an understatement," she told me before we went to Israel. "I still have trouble believing the sequence of events."

Unreich told Schiller that he wanted to make a documentary about the connection between survivors and their children. Schiller had explained that her father would never talk—that it would be a silent movie. Unreich was undeterred, and said that he was planning a trip to Israel, where he grew up, and would be grateful for the chance to film Daniela as she tried to engage with her father. Her father had no objection, so she agreed. "I told him not to worry, they were aware that he would say nothing."

Unreich had brought a young cinematographer who was born in Ukraine; he quickly established a rapport with Sigmund Schiller. His daughter asked him once again, for the film, to discuss his memories. He declined. "So we sat there in silence for a while, and I was happy that Liron was there to capture one of our 'conversations.'" Then the silence ended.

"I was eleven when my little sister was born," her father said, speaking Hebrew in a flat monotone, but with tears in his eyes. "I was very attached to her, and she was closer to me than to my mom. I taught her how to walk. Her first words, her first laughter was with me. I am the one who raised her."

Schiller had never before mentioned a sister, even to his wife. Daniela fought back tears in the film, and was fighting them back again in the family's living room. Her mother, who watched the film in silence, said in a whisper, "I never knew about your sister."

"She never grew up," the elder Schiller told Unreich in the film. "She was amazing."

Unreich asked if he remembered the last time he saw her.

"I remember the last time I didn't see her," he replied. "We had a maid who loved the kids very much, but she lived in a different village. When we ran away to hide in the forest, my mom took my sister to her house . . . After a while, there was a rumor that she had been executed. Two policemen came and took her to the fields. One was a 'humanitarian.' He didn't want her to suffer, so he took her toy, threw it away, and said go pick it up. That way he could shoot her in the back without her knowing."

In the living room, surrounded by book-lined shelves and bright pictures of birds, painted over the years by Yaffa Schiller, we all sat stunned, in silence. Before we left New York, Daniela Schiller had told me that her father finds being called a Holocaust survivor demeaning. "When people talk about the Holocaust, they talk about gas chambers, Auschwitz—the Holocaust is not just about that," she said. "It's about the little humiliations, the loss of dignity."

Her father made much the same point in the film. "People talk about 'Sophie's choice' as if it were a rare event," he said. "It wasn't. Everybody had to make Sophie's choice—all of us. My mother left behind a four-year-old with the maid. You don't think I was beaten and shot at? There are no violins in my story. It is the most common thing that happened."

Nobody moved in the Schillers' living room while the film continued. At times Daniela hid her eyes with her hands, and so did her father. For the most part, they were immobile. On camera, she asked him if he had consciously suppressed this information.

"Yes," he said. "You must suppress. Without suppression I wouldn't live."

"I have learned in my research," she told him, "that it should be the other way around. I think it's good to cry—you should bring back memories and relive them. And since you are not in the war anymore, it might be a good experience." At that, Sigmund Schiller shook his head and stopped speaking.

It's not clear if the experience has altered his memory of those events. But it has transformed his daughter's memories of him, and of her own life. She told me that she realized memory is "what you are now, not what you think you were in the past. When you change the story you created, you change your life. I created the story and brought these memories together, and now my past is different from the past I had before. Especially the memories of my father. He was a reserved man, a nontalking person. I know that, but there was this window when he was a different person. A very brief window, but now that is the person who inhabits each of my previous memories."

She sat quietly for a long time. Then she continued. "I picture him from the time I was in kindergarten. But now I only can see him with all the insight I have gained. My memory has been updated. I have spent much of my life trying to find a way to reconsolidate my father's memories, and ended up reconsolidating my own."

MEERA SUBRAMANIAN

The City and the Sea

FROM *Orion*

TWENTY YEARS BEFORE Hurricane Sandy slammed into the slim spit of land that is New York City's Rockaways, local artist Richard George was out planting trees. He was in his 40s then, and had shifted his home a few years earlier from Corona, Queens, to a 1920s bungalow colony in the Far Rockaways, abutting the Atlantic Ocean. He didn't know anything about trees, had never given a thought to dune ecology or sea surges, but he'd joined the board of the local Beachside Bungalow Preservation Association, and a friend gave them $15,000. The directive was to plant trees, so that's what he did.

"He planted the money in my hand," George recalls when I meet him at his cottage, a bright white bungalow with turquoise trim that matches his T-shirt. "I said, 'Where am I gonna plant trees?'" Then the artist saw the wide expanse of beach down the street, like a blank canvas in waiting.

George is convivial, a man who enjoys talking, which he's doing at a rapid pace in a heavy New York accent as we head from the cottage toward the beach. It's a clear spring day, making it hard to imagine the devastation that Sandy wrought when she landed on the shores of New York City, generating a 14-foot storm surge that dumped the Atlantic Ocean into thousands of homes, decommissioned trains, caused a Con Edison power station to explode, plummeted tracts of the city into darkness and others into fire, and took 43 lives. The storm cost an estimated $19 billion in damage and an unquantifiable amount of grief.

The street ends in a ramp that leads to an elevated boardwalk, rising a dozen feet above street level. In 1992, George tells me, you could walk right under the boardwalk with room to spare. Now, as we walk up the ramp to get to the length of the boardwalk, which was untouched by Sandy, we are surrounded on both sides by a dune so high that it's packed against the bottom of the boards. It's thick with plants, a seaside forest filled with bayberry, beach plum, autumn olive, wild rose, and Japanese black pine 15 feet tall. The first fiery hints of poison ivy inch out of the sand at toe level. This beach forest is what is considered a secondary dune; on the other side of the boardwalk, waves of beach grass cover the primary dune before sand takes over. The water appears so innocent, softly lapping onto the beach, sparkling in the sun.

"At first we just planted tiny little beach grass," George tells me. He gathered up 35 volunteers, and they spent a couple of weekends planting the starts. "One blade, one blade, one blade," he explains, his fingers poking into the air. Later they planted trees.

In 1994 they got another $15,000 and planted more. This time they rented a machine to dig the holes, and a neighbor cooked the volunteers lunch. Once the plants became established—NYC GreenThumb, which supports the city's network of community gardens, helped keep them watered through that first vulnerable summer—residents and beach bums could sit back and watch the sand grow. And grow.

"The beach grass grew by itself," he says as he bends to show me the thick, matted root system, exposed at the edge where Sandy's storm surge gnawed away half of the primary dune. He'd been standing in this very spot just five hours before the surge hit. The double-dune system that stretches for a few blocks on either side of George's street seemed to help fend off Sandy's deluge: the water breached on Beach 27th Street, where the dunes stop.

"The dunes were twice as high before the storm," he says, looking at the remains of the sacrificial sands. "It saved us."

Once upon a time, about 2 million years ago, the Pleistocene era locked up the world's water in glaciers miles thick. Then it warmed. It was about 10,000 years ago when the water of the melting glaciers was released to reshape the world into the coastlines we now associate with modern-day maps.

By all indications, though, the shape of those coastlines is about to change.

The archipelago of New York City's five boroughs has almost 600 miles of littoral zone between solid ground and watery sea, a place of straits and river mouths, bays and beachy backshores. It's also a place whose contours have been radically transformed by its citizens. A large percentage of the city's edges were created artificially, filled in and built upon with the false confidence that land taken from the sea is permanently allocated for terrestrial use. But on October 29, 2012, the record-breaking storm surge that swept over New York City flooded 51 square miles of that falsely allocated land—and like a finger from a watery grave, the high-water mark traced the coastline that once was and may soon be again. The mayor's office says that by the 2050s, 800,000 New Yorkers will live in 100-year flood plains, double the current number.

In 2008, Mayor Bloomberg convened the New York City Panel on Climate Change (NPCC), a multidisciplinary group comprising academic and private-sector experts on everything from climatology to oceanography to law and insurance risk. Their task was to assess how climate change might affect the city and make suggestions about how to mitigate those effects. Sea levels vary around the globe, and models still vary wildly, but the NPCC settled on 48 inches—the average height of a seven-year-old child—as the informal figure for estimated sea level rise in the New York City area by the 2080s. They readily admit that this number is subject to change, especially if the polar ice sheets get caught in a feedback loop that speeds up the process.

Given the huge number of people who live on the world's coasts, how will human populations—whether in Brooklyn or Bangladesh, Miami or Mumbai—adapt to an increasingly aquatic world? Do we stand strong and demonstrate our clever technical ingenuity with multibillion-dollar floodgates and waterproof buildings? Or do we humbly bid a hasty retreat, scrambling for higher ground while there's still time, waving a white flag to Mother Nature?

It seems increasingly clear that there may be a third way: an approach that blends a trace of conciliation with an abundance of creativity, using hints from the ecological past to design the coastlines of the future—and it could be the key to surviving in coastal communities in an age of rising seas. The way forward will require an unlikely collaboration, not just between public institutions and citizens, but also between humans and the one player too often left outside, literally and figuratively: nature herself.

When it comes to prospects for human life within the increasing reach of the ocean, design answers fall into two categories, the soft and the hard, the yin and the yang. Dunes, wetlands, and oyster reefs fall on the soft side; hard is the stuff I know from my childhood on the Jersey Shore: tar-smeared bulkheads and jetties jutting out at perpendicular angles in an effort to stanch the natural movement of sand. A floodgate is as hard as it gets.

Speculation about the construction of a great surge barrier surfaced soon after Sandy. The largest proposed gate would run parallel to the expansive Verrazano-Narrows Bridge, nearly a mile long. Outright critics spoke of engineering cockiness, ethical and legal conundrums, disrupted harbor ecosystems, and a false sense of security. More cautious skeptics noted that comparisons to countries that do employ floodgates, like the Netherlands, fall short when one considers that the eastern United States routinely weathers storms the Dutch can't imagine in their mild climate.

But instead of seeking out the big fix, most adaptation efforts are opting for a multifaceted approach. After Sandy, Mayor Bloomberg formed the Special Initiative for Rebuilding and Resiliency, and in June 2013 it released a comprehensive 440-page report, "A Stronger, More Resilient New York." In it the city proposes more than 250 different initiatives that would strengthen everything from the energy grid to communications networks to transportation systems. A proposal for a great flood barrier is notably absent.

The city's response echoes what is rising up like wrack on the high-tide line, a medley of ideas from all sectors across the boroughs and beyond. Many seek to embrace the best of technological advances and apply them in ways that foster, instead of resist, the fundamental laws of the natural world: Let the wetlands be wetlands, bird-festooned sponges. Remember the shape of New York's native coastlines. Cultivate sand dunes and beach forests. If Richard George and the Beachside Bungalow Preservation Association could save a bit of coastline with 35 volunteers and a neighborhood cook, imagine what a whole country could do if it acted with nature in mind.

The U.S. has over 12,000 miles of coastline, home to 53 percent of Americans. What can the rest of us learn about coastal infrastructure from the shorelines of America's largest city? I'd come to the Big Apple to find out.

It is springtime when I visit the western end of the Road to Nowhere, where the legendary NYC Parks Commissioner Robert Moses, a ruthless mid-20th-century urban planner who favored parkways over parks, left his indelible mark on the Rockaways in Queens. Of the five-and-a-half-mile-long boardwalk that ran along the Rockaways beachfront, more than half was destroyed by Sandy, and afterward the city spent over $140 million to rebuild beaches. The work continues today. I see one man scrape his shovel across a basketball court, fill his wheelbarrow with sand, and walk a half-dozen steps away to dump the contents on the beach; nearby, bulldozers move between the skeletal remains of the boardwalk's concrete stanchions. Annually dredges and offshore rigs vacuum up sand that is hauled onto the beach, where heavy machinery sculpts it into a shoreline in a continuous process called "nourishment."

I turn my back to the men and take in the two-mile length of Moses's Shore Front Parkway. In 1939, Moses orchestrated the divided four-lane parkway along the southern shore that was supposed to become part of a 90-mile road system along Long Island, linking the Hamptons in the east to Brooklyn in the west. (The project never came to fruition, hence the local nickname, Road to Nowhere.) But while Moses razed neighborhoods and built highways, a new generation of thinkers hopes to reshape the city with a gentler hand than the one he wielded. The Shore Front Parkway could be where it begins.

Watching the bulldozers do their work, I envision the space transformed into a system of forested beach dunes—it would be reclamation of the grandest sort. This is the idea of Walter Meyer and Jennifer Bolstad, the husband-and-wife team of Local Office Landscape & Urban Design, on-and-off Rockaway residents, and avid surfers. In the immediate aftermath of the storm, Meyer worked with others to form Power Rockaways Resilience, which delivered hand-built, shopping-cart-sized solar generators to the hardest-hit parts of the Rockaways so storm survivors could charge cell phones, laptops, and small power tools. The White House recognized him as a Champion of Change.

Now Local Office is looking long-term. They have set their sights on transforming the lingering Moses legacy of the Shore Front Parkway into a double-dune forest system, similar to what Richard George showed me on a small scale on the Rockaways' eastern end. They propose downsizing the parkway from its current 80-foot width to a reasonable 30-foot road and using the new space to develop dunes planted with trees and shrubs that provide beachfront storm surge protection. The boardwalk would continue to exist, nestled between the primary and secondary dunes. They imagine a system like this stretching across the entire 15-mile length of the sea-facing Rockaways, a natural double-duty sea wall of sorts.

"This is a story about trees," Meyer tells me. "It's less about the dunes than what the dunes support, this coastal forest. It's a living armor." Globally, Meyer says that more and more nations are turning to softer approaches to dealing with sea level rise. "The Dutch don't build dykes anymore, they're building sandbars and sand dunes," he says. "The Japanese are building forests. The Australians are building reefs."

By late 2013 the NYC Department of Parks and Recreation had commissioned Local Office for drawings, the first step for a pilot project that may bring the double-dune forest vision to life. Meanwhile Meyer and Bolstad continue their community-based efforts, teaming up with reTREEt America, which plants in areas struck by disaster, to bring volunteers together to landscape with native plants on property where dunes are being developed. The most recent planting day evolved into a block party, where volunteers were plied with fresh fish and beer as homeowners gave their thanks.

"There are some parcels that Mother Nature owns," New York governor Andrew Cuomo said in his State of the State address a few months after Sandy. "She may only visit once every few years, but she owns the parcel and when she comes to visit, she *visits.*" Nature —in the form of rising seas—has already begun to visit a stretch of the East Coast from Cape Hatteras to Boston, which is experiencing sea level rise greater than the global average. And in 2013 the New York–area flood zone doubled when FEMA released its new maps. Some scientists think that even this is a grave underestimate of coastlines at risk.

A bit of surrender, then, may be unavoidable. For Meyer and Bolstad's double-dune forest to exist, for example, some would lose their ocean views and others perhaps their houses, houses that are so close to the ocean that there is no room to create a dune system. But the reality is that many of those houses were already relinquished, swept off their foundations during the storm in a sudden unplanned event that is the opposite of managed retreat.

The crooked flagpole that leans in the sand at the end of Kissam Avenue in Staten Island's Oakwood Beach neighborhood is a reminder that a dune can only be battered so many times before it's breached, and with rising sea levels, the concern is real. Most of the Fox Beach area of Oakwood has signed up for a buyout—at prestorm home values—that Governor Cuomo proposed post-Sandy. The offer was impossible to pass up for many beat-up shoreline dwellers who lived in what was essentially a bathtub, vulnerable to storm flooding from the west, ocean surges from the east, and brush fires from the phragmites-filled wetlands that close in from the north and south. The whispering fronds betray a neighborhood that was once, and remains, a wetland.

"The storm would've killed me if I'd been home," says resident Bill Bye, whose single-story bungalow on Kissam Avenue filled nearly to the ceiling with seawater during Sandy. When I meet him, Bill sits in a plastic lawn chair in front of his home of 30 years with a single sheet of paper in his left hand and a cigarette in his right. The paper is a New York City assessor sheet, which shows the value of his home before October 29, 2012. He hopes the assessor will agree with the value and that the government will issue a check—soon—allowing him to move on with his life. "It's gonna happen again," he says. "It's too dangerous to live here."

But Oakwood Beach and its wetlands don't have to signal disaster in waiting. The area is also part of a growing system known as the Staten Island Bluebelt. Run by the NYC Department of Environmental Protection, the Bluebelt is the aquatic equivalent of a greenbelt. The extensive system of engineered ponds, creeks, and wetlands naturally provides flood control by absorbing storm surges across 10,000 acres and 16 watersheds that span the southern end of the borough, Oakwood Beach among them.

Yet the word *engineered* evokes the wrong visual. Instead, imagine Staten Island circa 1850, with babbling brooks passing under stone bridges and along banks blooming in purple-flowered pickerelweed. Bluebelt designers, who use words like *beautification* and *countrify* to describe their work, draw their inspiration from mid-19th-century photographs and native plant stock as well as the latest knowledge of hydrological dynamics: strategically placed riffles and pools along with abundant plantings help soak up storm runoff pollutants like nitrogen and phosphorous. A moratorium on new building in wetlands along with buyouts like the one in Oakwood Beach are helping to expand the Bluebelt's reach. "This is what some people like to call 'thinking out of the box,'" says Staten Island borough president James Molinaro. "Instead of putting down sewers, you use nature to purify and disperse storm water. There's both beauty and effectiveness."

The Bluebelt costs considerably less than a typical underground sewer system, and it provides the added benefit of fostering community open spaces and wildlife habitat. Neighborhoods can "adopt" a piece of the Bluebelt, and volunteer groups help with cleanups. The result is an ecosystem-wide maintenance program that brings together citizens and government in a creative way that is winning awards.

Another unlikely but increasingly sensible partnership for a livable New York is between humans and a more aquatically sophisticated member of the natural world—oysters. The city's waterways were once dense with them. Jamaica Bay alone, a vast network of wetlands and marshes on the north side of the Rockaways, used to send 300,000 bushels of oysters to city markets each year. But by 1921 the impacts of sewage and industrial effluent, overharvesting, and dredging had taken a toll, and the oyster beds perished or, for health reasons, were no longer harvested.

New York City waters are now cleaner than they've been in decades, and spat—the little larvae of oysters—are floating about, seeking out some substrate to latch on to so they can grow.

"There *are* oyster larvae," says landscape architect Kate Orff. "There's just no place for them to land. So there are these babies, but then . . ." She pauses and makes a long downward whistle indicating failure. "I get depressed about this, as a mom of two."

Orff has proposed bringing back New York City's lost oyster reefs—which could naturally help shield the city from future storm surges—in a project called Oyster-tecture. We meet at a round table in the kitchen of her SCAPE Studio offices on Lower Broadway, with a foggy morning view across Manhattan to the Hudson River.

Her idea is to make a home for the oyster orphans, starting in Brooklyn's Gowanus Canal, one of the most polluted waterways of the city. Piers are already in place; cages will be lowered into the water that are filled with fuzzy rope, providing lots of surface area for spat to land. (The cages are to keep hungry New Yorkers from prematurely harvesting oysters that won't be suitable for human consumption for years to come.) She was thrilled when she heard Governor Cuomo give a shout-out to the bivalves after the storm: "It came out of his mouth—'We need to be thinking about oyster beds'—and I'm like, yes, that's the goal!"

Orff advocates dredging and filling waterways in a way that supports underwater ecosystems instead of destroying them. "I think a big part of this is thinking about it holistically," she says. "You can use dredging and filling in a cut-and-build concept throughout the harbor to create a new set of edges and cross-sections that can become armatures for habitat." As the oysters build upon themselves, making reefs out of their own shells, they work continually to clean the waters, acting as natural filtration systems.

To consider the return of oyster reefs to New York is to explore the bathymetry, or underwater topography, of the waters that surround the city's islands. Orff explains the mechanics, but it's as much a history lesson.

"There was once this historic 3-D mosaic of underwater bathymetry, which included barrier islands and oyster reefs and shallower shoal areas that make a threshold into the inner harbor," she says, using a marker to show how the stretch of Sandy Hook on the Jersey Shore wants to reach up and connect with Coney Island in a series of channels and shoals. If the landmass was permitted to form again, it would help diminish the impact of sea surges. One hard passageway could allow a steady shipping channel. "See this signature?" She taps her marker on the area between Staten Island, Brooklyn, and New Jersey. "This whole area was once shallows, back in 1766, fortified with layers of ecological systems."

Orff isn't the only person in the city thinking about oysters. Marit Larson, the NYC Parks Department's director of wetlands, is working on oyster restoration on the north side of the city. She says that increasingly oysters have been showing up on tires and concrete rubble in the Bronx River. In 2006 the Parks Department started designing artificial reefs to supplement the river's substrate; now they're working with partners that include NY/NJ Baykeeper and New York Harbor School, which trains low-income kids in hands-on maritime stewardship as they dump 120-ton loads of shells into an area where the East and Bronx rivers converge. One survey taken just before Sandy struck found more than 11,000 oysters along a mile-wide swath of water. Last summer students relocated another 100,000 farm-raised spat to the growing reef.

When referencing historical patterns, Kate Orff refers repeatedly to something called the Ratzer Map. In 1766, 10 years before the rebellious colonists unshackled their new nation from the British empire, cartographer Bernard Ratzer was trolling the New York Harbor, recording its curves and depths. His sepia-toned map now hangs at the Brooklyn Historical Society, a memory of the city's once-flourishing natural infrastructure that, if brought back to life, could help mitigate the type of damage experienced during an event like Sandy.

Orff's pen sweeps across her drawing. "This is a diagram of an ideal world of nature's protectors. You have layers and layers of barrier dunes, and then vast tracts of wetlands. Wetlands don't make sense if you have a building and only a little wetland. Wetlands only make sense if they are extensive, so they can really dissipate wave energy."

Of course, efforts like these can only do so much to counteract the effects of climate change. But if combined with a comprehensive rebuilding of New York's natural features, they could make a real impact. "There is literally nothing that could have stopped the Sandy surge," Orff says. "But hard and soft infrastructure solutions could be combined in a win-win-win scenario that would revitalize the harbor landscape, clean the water, and begin to address coastal protection."

The Ratzer Map holds some clues to what a more resilient New York might look like, but it's a static snapshot compared to the Mannahatta Project, which documents the changes to Manhattan island that have occurred since that fateful year of 1609, when the explorer Henry Hudson first sailed his vessel, the *Half Moon,* into New York Bay. The project, now extending to all five boroughs and renamed the Welikia Project, is the work of Eric Sanderson, senior conservation ecologist with the Wildlife Conservation Society. Sanderson spent more than a decade piecing together the city's original ecology, mapping where its creeks once flowed, what its shoreline once looked like, and which flora and fauna filled the ecosystem. The results are a roadmap for designers seeking to know what was, in order to guide what will be. Because there is no indication that we humans will stop tinkering with our environment, the hope is that the adaptations we make embrace the third-way approach that balances human needs and ecological limitations.

"Constructing human habitats that work with nature rather than against it will have the greatest benefit for people and nature," Sanderson says. "Probably the most important work to be done, but also the hardest, is adapting the social, legal, and economic institutions to incorporate messages from the natural world."

Even with those systematic challenges, Sanderson is solid in his belief that we can learn—and benefit—from listening to what the natural world is telling us.

"Hurricane Sandy was information encoded in a storm," he says. "If people begin to see the nature of our place, then they can begin to see the landscape strategies that history suggests are protective and adaptive over the long run: specifically, a combination of beaches, dunes, and salt marshes."

The process of reimagining New York City's infrastructure with climate change in mind was underway before Sandy, but the storm's devastation underscored the urgency of learning from nature and then planning and designing with her machinations in mind. There will be some managed retreat—some withdrawal from coastal areas that one hopes will be graceful—but there are also ways to stay along our shorelines safely. It demands rethinking the meaning of edge, redefining it as something more fluid than the single hard line conveyed by a cartographer's pen. To get there will require an era of collaboration and partnership, from government-level climate change panels to grassroots citizen efforts, from design competitions to smartphone apps to gatherings of engineers, city planners, and scientists. It will take people stepping out to meet their neighbors—before the high waters come.

Ultimately such a collaborative approach could be our generation's grand act of conciliation with the changing forces of the natural world—one that could represent a cautious step into a future that will allow us to keep some of our coveted seaside haunts while also conceding that some places we've set up camp are simply not ours to inhabit.

KIM TODD

Curious

FROM *River Teeth*

FLAT HEAD, LIDLESS eyes, body dirt brown, the Surinam toad slithers through the pond like animated mud, an amphibian golem. Long fingers filter the swamp floor, sweeping fish and worms into a tongueless mouth. Some romantic gave it the alternate title "star-fingered toad," because after each foot divides into toes, the toes divide again, creating a "star" at the tip. But celestial the toads are not.

The most disturbing part, though, is the way they breed. Mating, they somersault through the water, the female dropping eggs on the male's belly when the pair is upside down. Turning, he brushes them onto her back and fertilizes. The eggs sink into her body, leaving pale warty bumps until, months later, fully formed miniature toads (not tadpoles) burst out and swim away. When they've gone, her back is all holes like a drained honeycomb.

In the 17th century, every collector in Amsterdam coveted a Surinam toad, in a box, in a painting, in a jar. The Dutch had colonized Surinam, a small country on the northeast shoulder of South America. Between planting sugar and pineapple, they marveled at the creations of the tropical rainforest. Frederik Ruysch, botanist and midwifery expert, kept one in his cabinet of curiosity, a gathering of oddities, the precursor to museums, where his friends could come gawk. Nicholaes Witsen, burgomaster and avid collector, reported seeing the disconcerting toad and described it: "All the back is open as a wound."

Some viewers interpreted the ugly appearance as a judgment

for ugly habits: "The pipa is, in form, more hideous than even the common toad; nature seeming to have marked all these strange mannered animals with peculiar deformity," wrote Oliver Goldsmith in *A History of Earth and Animated Nature* in 1774.

Those who seek her out today still have a visceral response. Comments on a YouTube video of toadlets tearing the skin and kicking free include "damn nature you scary!" and "as a lover of animals im ashamed to say i said KILL it with fire and dont stop . . . kill it burn it kill it burn it! then bury it and shoot the grave!" (A more sober-minded commenter suggested that compared to watching a human birth, it wasn't bad at all.)

Given the strength of the recoil, why can't we look away? The YouTube video has been watched almost half a million times. Perhaps now you are tempted. Knowledge of South American toad habits has little practical application. What is the nature of the itch we call "curiosity"?

Like the toad, curiosity is a strange beast. The investigating mind moves like a sleek little mammal, a mink maybe, rubbing up against things in the dark, trying to determine their shape, occasionally ripping with sharp teeth and pawing through the opening. Or perhaps a spider, creeping precisely, attaching silk here, and here, and here to impose a pattern where before was just air.

Curiosity can be as obsessive as hunger or lechery, swamping the senses. But it is notoriously fickle too, slinking away as soon as it is satisfied. Its subjects seem so frivolous: a baby giraffe, a dodo skeleton, the Surinam toad. George Loewenstein, author of *The Psychology of Curiosity,* summed it up: "The theoretical puzzle posed by curiosity is why people are so strongly attracted to information that, by the definition of curiosity, confers no extrinsic benefit." Saint Augustine defined it as "ocular lust," the desire to stare at an object, animal, or scene and let the mind roam. Charles Darwin, who you might think would value the trait, saw it as the enemy of substantive inquiry. He wrote in a letter to a friend: "Physiological experiment on animals is justifiable for real investigation, but not for mere damnable and detestable curiosity."

This kinship to physical passion—the strength of desire, the burst of delight—makes curiosity waver between vice and virtue. Intellectual curiosity sparks science, art, all kinds of innovation. Here, in most of 21st-century North America, it is held in the highest esteem. For much of history, though, coveting the secrets of the world and mulling over mushrooms and vipers threatened to drag one from thoughts of God. Religious and worldly contemplation were at odds. Thomas Brooks, an English nonconformist preacher of the early 1600s, warned, "Curiosity is the spiritual adultery of the soul. Curiosity is spiritual drunkenness."

Don't unlock the door. Don't open the box. Don't eat the apple. Fairy tales, Greek myths, and biblical stories caution against giving curiosity free rein. The warnings are dire. But so often, like Pandora, Eve, and Bluebeard's wife, we still extend our hand.

In the late 1920s, Henry Nissen, a slight, bespectacled psychology graduate student at Columbia, took a more scientific approach to the question of the nature of curiosity. He noticed that white rats explored their cages, even if they had to delay eating or mating. How far would they go to satisfy their curiosity? he wondered. Nissen put male albinos in a cage linked to a passageway that would give the rats an electric shock when they crossed it. On the other side of the passageway, he created an irresistibly interesting environment for a rat (he hoped): a mini-maze of sheet-metal walls in a pine base, seeded with corks and piles of wood shavings and a rubber mat. The rats investigated, even at the price of pain.

Of course, this curiosity is not merely idle. It could be a life-or-death matter for a rat to know what lurks around an unfamiliar corner. But for humans it's more complex. Why, for example, might anyone want to read about rats shocking themselves in pursuit of wood shavings?

These questions intrigued Daniel Berlyne, working on his PhD in the 1940s at Yale. He distinguished between the exploring curiosity, shared with animals, and the kind he felt was uniquely human, "epistemic": that curiosity based in the desire for knowledge. What type of things triggered human curiosity? he wondered. What would be our equivalent to corks and a rubber mat? To find out, he gave undergraduates in the course "Normal Human Personality" at Brooklyn College a list of 48 questions about invertebrates.

For example:

"What crops do some ants cultivate in underground 'farms'?"

"What form does the clam's brain take?"

"How does the spider avoid being caught in its own web?"

"How do sea wasps swim?"

Now, he said to them, and I say to you, which makes you most curious?

Sifting through his results, Berlyne concluded that curiosity is spurred by the novel, the complex, the ambiguous, the uncertain, and the surprising. When elements previously thought of as incompatible are harnessed together—a juxtaposition—the curiosity grows stronger. Loewenstein described these traits as "violated expectations," and noted that often, the closer the subject matter was to the observer's life, the more intense the need was to stare, to figure it out.

In a study published in 2009, scientist Min Jeong Kang and others recreated Berlyne's experiment using MRI equipment that let them see where blood flowed in the brain. They showed Cal Tech students slides with questions, then asked them to rate how curious they were about the answers. (The questions ranged farther afield than Berlyne's, covering rock bands, politics, and snack foods. It is worth noting, though, that one of the questions that prompted the most curiosity was "What is the only type of animal besides a human that can get a sunburn?" Animals make us endlessly curious.) These intellectual questions spurred a physiological response. Curious students' pupils dilated. Activity increased in the caudate nucleus, the bilateral prefrontal cortex, and the parahippocampal gyri, which the scientists interpreted as the brain anticipating a reward. Other studies have charted the way curiosity triggers the production and release of dopamine, a neurotransmitter associated with heightened arousal and motivation, creating the experience of pleasure. The mind becomes receptive, wakes up; that little mink grows hungry.

Incidentally, reading Berlyne made me realize how similar these traits are to what I ask of my writing students. The opening of their essays should make an irresistible offer to the reader—to show her that her assumptions about the way the world works are wrong, to present an odd juxtaposition and promise to unravel it, to reveal the strange hidden at the heart of the familiar. Not surprisingly, Berlyne later went on to study aesthetics, exploring what makes viewers stare at one painting longer, and with more pleasure, than another.

Given Berlyne's formulation, it makes sense that the Dutch couldn't stop looking at the Surinam toad. The animal, its life history, complicated questions they were asking about humans. For example, how did we breed? Some claimed they could see the man in miniature inside the sperm head. Others felt the egg contained the whole, a preformed creature that started to grow when nudged by the sperm. Still others argued for a more gradual development. And here was a swarm of tiny toads, emerging complete from their mother. It violated expectations of toad reproduction (eggs) and mammal reproduction (live birth). And from the mother's back? What could that mean? It was a disturbed mirror. Irresistible.

One of those most fascinated by the Surinam toad during Golden Age Amsterdam was an artist and naturalist named Maria Sibylla Merian. She was curious about metamorphosis—of insects, of frogs—and she visited the best Dutch cabinets of curiosity, including those of Ruysch and Witsen. But she was dissatisfied by what she found there. The preserved specimens didn't tell her how the creatures lived: what they ate, where they hid, how they moved from shape to shape.

So she went to Surinam to investigate, spending two years gathering plants and moths and lizards, raising them through their life cycles, taking notes. For her painting of the toad, she captured a female and dropped it in a jar of brandy as soon as the young broke the skin. In the final image, the Surinam toad is miderup-tion. The mother's face is impassive. One little toad swims behind her, ready to dart off into its own perverse life.

Several years ago my own curiosity took me to Surinam. I was writing about Maria Sibylla Merian and her thirst to understand the insects, lizards, fruits, and toads of South America. I hoped to retrace her footsteps, wondering about her drive and self-conception. What was she doing there? How did she get away with it in a time when the slightest sign of eccentricity could cause someone to be burned as a witch? That a woman would travel to Surinam with her 21-year-old daughter to conduct field studies 300 years ago violated expectations. That strange juxtaposition of person and time and interest made me itch.

It was not an easy journey, for reasons both practical and emotional. The first airline where I bought a ticket went bankrupt a week before the plane left the ground, no one in Surinam answered my phone calls or e-mails as I tried to make arrangements, and I had to leave my nascent family—my husband and 10-month-old twins—for a week and a half. But this was my work; the money from this book was sustaining us, I told myself. Besides, I wanted to go. Chasing an epistemic question dilated my pupils, lit me on fire.

Despite my itchiness, though, I was not very brave. Not as brave as Cheryl, the anthropologist who hitchhiked on canoes into the rainforest to interview remote tribes of Maroons, descendants of escaped slaves who had fled the Dutch sugar plantations. She looked at the forest floor, found evidence of earlier ways of life. Cheryl said she'd arrange for me to fly to her study sites and I said no, afraid of her descriptions of the drug dealers and prostitutes and miners in the interior, of evenings spent picking ticks off each other over a sloth stew. Who knew what might happen? All I could picture was the tiny plane going down and my death in flames and broken branches.

Instead I took the bus from the capital of Paramaribo to the National Zoological Collection of Suriname at Anton de Kom University. The university logo displays a Surinam toad next to a microscope. After a quick hard rain, steam rose off my clothes as the curator of invertebrates gave me the tour, highlighting the species that enthralled the Dutch: the lantern flies with heads reputed to light up, bat skeletons, hundreds of kinds of beetles. In a hallway packed with specimens, a Surinam toad glowed in a jar.

Leached of color, floating in space, unmoored from time, the toad had transcended its lifespan with the help of pickling. She was less golem than ghost. The ragged holes left her looking mutilated. I could see that the attraction to her is snarled up in fear: dread of infestation and disfigurement, loss of bodily integrity.

Looking at the suspended skin and bone made me think for a moment, not about the toad but about the jar. It's hard to analyze—a form of dissection—subjects in motion. How much easier to go part by part when they are held still. Think of where we put our curiosities: in a bottle, in a zoo, in a glass display case with their stuffed peers. Faced with an object in isolation, the mind can stretch its muscles, crawl and explore. The blank background forces focus, a situation that we crave but that is ultimately artificial, stripped of vital information.

The image of the pale toad in the jar makes me think of Barbara Benedict's recent book *Curiosity: A Cultural History of Early Modern Inquiry*. She traces the evolution of curiosity during the 17th and 18th centuries in Europe, the era of Maria Sibylla Merian and many other curious folks. This period included the scientific revolution, Linnaeus's categorization, comprehensive exploration of the Americas, the birth of the novel. Even as curiosity gained respect, though, it remained suspect in women particularly, a category that includes Eve, Bluebeard's wife, and Pandora as well as gossips, women thrilling with useless information. The reason? As Benedict says, "Curiosity is the mark of discontent" and "curiosity is seeing your way out of your place." One of the things that makes us most curious is the suggestion that the world isn't how we think it is, that our categories are the wrong ones, and the promise is that the answer to our questions will give us a different, fuller, better view.

It's easy to dismiss this criticism as reflecting another world, one more obsessed with maintaining the status quo and keeping women in their place. But Benedict makes other points about the nature of curiosity, points that are harder to ignore. The way, for example, that items in the cabinets of curiosity (and modern-day museums and zoos) have been stripped of their use, of their ability to work. A bow and arrow, a ceremonial mask—to be a curiosity, they have to become not a way to feed yourself or talk with the gods but an artifact. To own one means you have the status to possess a bow and arrow for decorative purposes. Curiosity can turn the objects of its desire into, well, objects. They can enter the system of commerce; they can be used and then discarded. Bluebeard, Benedict reminds us, was also curious. He had his own collection—of women's heads, which he could study without having to contemplate their subjectivity, their humanity.

Before I left to go to Surinam, the toad seemed to be a story about ocular lust and the extent people will go to satisfy it. But it's also a story about the glass, its insulating properties, the attraction of those boundaries. They keep us safe—not just safe to make precise observations, but also safe from being observed ourselves. Maybe this is the real danger of curiosity, the teeth in the warning. Though it's easy to laugh at Brooks, the preacher who claimed "curiosity is spiritual drunkenness," maybe he has a point about the risk that it will distort our sense of what is important, that it will obscure the big picture.

For curiosity to have value, perhaps we have to allow it to be the beginning of something larger, to pursue it past the initial itch, the spark of hunger, the quick answer, the dopamine burst, to the "real investigation" Darwin asked for. Darwin, so hard on curiosity, was relentless at looking at the big picture, of looking at creatures through time, of challenging categories and turning the mirror back on humans. We have to see the disturbed world it implies, and ourselves living within it, moving beyond the role of observer, of questioning mind.

Several days after seeing the toad in the jar, I paced along one of the trails through the Brownsberg Nature Park, at the edge of the rainforest, through the tall trees with invisible crowns, the wasps building bulbous nests on the branches, the odd smashed fruit spilling its seeds in the dust. The plants Merian picked and painted often didn't have European names—she worked with the head of Amsterdam's botanical gardens to categorize them—but in places she used the Amerindian name or left the plant nameless. Brown-furred flanks rustled in the bushes. Monkey tails unfurled into upside-down question marks.

Maybe my question about Merian was never going to have a definitive answer. Certainly I could never know what was going on in her mind. Maybe the question was going to change as I pursued it. Certainly its pursuit was changing me.

Off to the side of the trail, a mud-colored head poked out of a small pond. I knelt to see if it was a Surinam toad, but with half its body submerged, no real line between skin and water, I couldn't say. No label declared it rare or important or special or forbidden. No brandy or videotape scissored it from time. It had escaped the box, like one of Pandora's demons, and now might breed or rot or pioneer some new behavior, do its work of living, participating in the fleshlife of the forest. And for the moment I was content not to know. Anything could happen. And then it ducked back under, out of reach.

Answers (just in case you were curious):

Ants cultivate edible fungus in underground "farms." Interestingly, the 17th-century Dutch generally and Merian specifically were also fascinated by these same ants, the leaf-cutter ants of Central and South America. They scissor bits of leaves and carry them over their heads like green sails back to the anthill. Early naturalists proposed many theories for what the ant was doing—using the leaves as food or miniature roofs—but it would take a 19th-century biologist to unravel the mystery.

The clam doesn't have what we would recognize as a brain. It has three "nerve centers," pairs of ganglia near the mouth, near the foot, and near the back of the body.

A spider doesn't stick to its own web because only some threads of the web are sticky, and the spider avoids them. In addition, its legs are covered with a nonstick coating, and specialized claws allow it to move lightly and precisely.

Despite a fascinating life history presented to the undergraduates by Berlyne in an effort to spur their interest, the nine-inch-long sea wasp, a creature with a fatal sting whose main enemy is the sea-wasp-eating porpoise, is fictional.

And, finally, the correct answer in the Cal Tech study was "pigs sunburn." More recent research has shown that whales sunburn too.

DAVID WOLMAN

The Aftershocks

FROM *Matter*

GIULIO SELVAGGI WAS asleep when the shaking started. It was the night of April 5, 2009, and the head of Italy's National Earthquake Center had worked late into the night in Rome before going home to crash.

From the motion of his bed, Selvaggi could tell the quake was big—but not close. When you're near the epicenter of a major quake, it's like being a kernel of corn inside a popcorn maker. When you're farther away, the movement is slower and steadier, back and forth, as the shock waves hit you.

Selvaggi hopped from the bed and checked his phone, but there were no messages. He hurried into the living room, dialing the office on the way.

"Where is it?" he asked.

"L'Aquila, 5.8," came the answer.

(It would later be classified as a 6.2.)

Selvaggi's first thought: At least it's not a 7. A magnitude 7 quake centered in L'Aquila, a medieval town high in the mountains, would have killed 10,000 people.

Seventy miles from Rome, Giustino Parisse had already been woken twice by tremors. The second one, at 12:39 in the morning, had stirred his whole family. Checking the house, Parisse, a 50-year-old journalist with the L'Aquila newspaper *Il Centro*, met his teenage son in the hallway.

"Questo terremoto ci ha rotto," said 17-year-old Domenico, restless. *This quake is breaking our balls.*

"I know, I know," Parisse replied. "But you have school tomorrow. You really have to go back to bed."

He switched on a light to peek in on his 15-year-old daughter, Maria Paola. She wasn't asleep.

"We're all going to die here," she said.

Startled, Parisse tried to muster a joke. "Nothing could ever kill *you,*" he said, and headed back to bed.

Three hours later Parisse and his wife woke to an avalanche of plaster and brick. They clawed and scrambled their way into the hall, lighting their path with a cell phone, and tried to reach the children. But it was too late: Domenico and Maria Paola were buried, dead.

The 28-second earthquake had demolished hundreds of buildings throughout L'Aquila. By the time the shaking was over, 297 people had been killed, more than 1,000 injured, and tens of thousands were made homeless.

During the winter and early spring of 2009, Selvaggi and other seismologists at Italy's National Institute of Geophysics and Volcanology had been monitoring numerous tremors around L'Aquila. The sequence of small quakes over a short period of time, known as a "seismic swarm," is distinct from the aftershocks that follow a big quake.

And in places like L'Aquila they are not necessarily abnormal. Local media repeatedly relayed that generic message to the public. Regional government officials insisted there was no need to fret, despite chronically unenforced building codes. The Civil Protection Department for Abruzzo, the region where L'Aquila is located, even issued a press release flatly proclaiming there would be no big earthquake.

But the people of L'Aquila were understandably concerned. Over the centuries the city had been devastated by several major quakes: one in 1703 killed 10,000 people, and a magnitude 7.0 quake in 1915 killed 30,000. This history has given rise to a culture of caution. When the ground seems especially temperamental, many residents—like their parents and grandparents before them—grab blankets and cigarettes and head outside to mill about in a piazza or a nearby park. Others sleep in their cars. Better not to be in an ancient building that hasn't been seismically reinforced.

As the swarm continued, anxieties were compounded by a local personality named Giampaolo Giuliani. Giuliani uses a homemade apparatus to try to predict imminent earthquakes. His proclamations—and the amplifying power of media interest in them—earned him a reputation in town. During church services at Santa Maria del Soccorso or over an orange soda at Bar Belvedere, he was often greeted not with *Buongiorno* but *Tutto a posto?* (Everything look okay?). One local news outlet referred to him as "the prophet of doom," and every time the earth shook that winter it seemed to validate Giuliani's incessant agitation.

By late March thousands of tremors had happened, dozens of them hitting 3.5 on the Richter scale. Then, on March 30, a 4.0 tremor catapulted the situation from tense to near madness. Sensing the need for a gesture that would calm the public's nerves, the country's Civil Protection Department—Italy's equivalent to FEMA—decided to call in the country's top experts, the Serious Risks Commission, to assess the situation. Selvaggi, the seismologist in Rome, wasn't on the commission. But his boss, Enzo Boschi, was. A titanic figure in the Italian science community, Boschi asked Selvaggi to come along and talk with the group.

Before the meeting Italy's civil protection chief, Guido Bertolaso, called the regional office in Abruzzo. According to a transcript that was later leaked to the media, Bertolaso said the goal of the meeting was "to shut up all of these morons and calm people down."

He was particularly annoyed that local officials had tried to counter Giuliani's claims with the preposterous response that there would be no quake.

"Don't say this business about 'No more tremors are expected.' That's a totally fucked-up thing to say," he said. "You can never say anything like that when talking about earthquakes . . . not even under torture."

Instead, he explained, he wanted a "media operation." Just wait: Italy's earthquake luminaries were on their way to L'Aquila. They would clean up the mess.

The meeting, attended by seven experts, including Selvaggi and Boschi and a handful of local officials, took just an hour and a half. Their conclusion: A major quake in the near term was unlikely. But remember, this is earthquake country: you never know. Boschi's words during the meeting would later prove pivotal. "A large earthquake along the lines of the 1703 event is improbable in the short term," he said, "but the possibility cannot definitively be excluded."

An Abruzzo official pressed the prediction question once more. "We would like to know if we have to believe those people who go around creating alarm." She was referring to the self-proclaimed expert, Giuliani. Such claims have no scientific basis, replied commission chair Franco Barberi. "The seismic sequence doesn't foretell anything, but it surely refocuses attention on the seismogenic zone where, sooner or later, a large earthquake will occur." The only thing you can do to protect people in such a place, he reminded her, is make sure structures are safe. As scientists and engineers repeat almost like a rosary: earthquakes don't kill people; buildings kill people.

Residents of the town learned of the group's assessment through news sound bites, including a snippet of an interview with commission member Bernardo De Bernardinis, whose background is in hydrology, not seismology. The interview was filmed before the meeting but broadcast afterward, giving the false impression that it was a summary. Was the swarm a sign of worse to come? "On the contrary," De Bernardinis said. "The scientific community assures me that the situation is good because of the continuous discharge of energy."

It was a wildly incorrect statement. Fault lines are not pressure valves, and small tremors don't necessarily release energy that otherwise would have contributed to some bigger earthquake. Virtually all seismologists agree that there's no known correlation —positive or negative—between the timing of small quakes and large ones. The television reporter, however, found De Bernardinis's answer to be so satisfying that he decided to finish on a jovial note: "So we should go and have a nice glass of wine?" Oh yes, De Bernardinis said, recommending a vintage from his hometown.

Six days later L'Aquila and so many lives were destroyed.

Many of those who were in L'Aquila that night are haunted by the sound of the quake, an eerie and inescapable rumble that felt as if a supernatural force were bearing down on them. Others recall the red cloud that quickly covered the city, caused by countless red terracotta roof tiles shaking, falling, and breaking.

Once the shock had subsided, bodies were removed, and rubble cleared, survivors began speaking out against the commission. They insisted that it had been the reassurances of the experts that persuaded people to stay inside the night of April 5, even after the two tremors, rather than head outdoors and away from unsafe buildings.

One critic of the commission was Parisse, the journalist who had lost his teenage children. "As a father," Parisse told me, "I'm the one ultimately tasked with keeping my children safe." But that does not negate what he sees as a failure of responsibility on the part of the scientists and engineers. "They came to L'Aquila to reassure people, not to assess the risk," Parisse says.

Many others felt the same. Soon a lawsuit was being put together by locals, accusing the scientists of negligence in their statutory duty to assess risk and advise how best to minimize loss of life and property. Parisse joined the suit. As a journalist he could write more articles, he says, but their impact would be minimal and ephemeral. "With the justice system, you have a chance for a more lasting truth."

The following spring the seven men—five commission members and two experts, including Selvaggi—were charged with manslaughter. The claim: in the March 31 meeting and statements afterward, they had knowingly neglected their responsibility to inform the population about the risk at hand.

The announcement of formal charges triggered a flood of condemnation around the globe. "Risk of litigation will discourage scientists and officials from advising their government or even working in the field of seismology and seismic risk assessment," declared an editorial in the American Geophysical Union journal, *Eos.* The CEO of the American Association for the Advancement of Science wrote to Italy's president, reminding him that a veritable mountain of research, much of it conducted by Italians, shows that earthquakes are unpredictable.

But Fabio Picuti, Abruzzo's 45-year-old public prosecutor, was either immune to the scorn or emboldened by it. Outsiders don't know what happened here, he argues. They don't know about Bertolaso's order for a "media operation." They can't comprehend the full impact of De Bernardinis's comment about "energy discharge." They're unfamiliar with our customs, our humility when it comes to Mother Nature's destructive force.

As he put it: "The easiest media line is to say that this was 'science on trial.'"

Abruzzo's main courthouse was so damaged by the earthquake that the trial was held in a blocky set of temporary buildings in an industrial part of town. It was the summer of 2011. On a gate outside someone had made a sign: "To punish those who killed our children is not revenge. It is a way to let our children die a little less."

Picuti's case against the scientists built a pattern: residents resisted their established habit of fleeing their homes during tremors because of an overly calming message from the distinguished commission. It was "disastrous reassurance," as Picuti likes to put it.

The trial was consumed by testimony from injured victims and the bereaved. People spoke of relatives who stashed blankets and cookies by the door to grab before exiting in the event of a tremor but had chosen to stay inside after seeing the TV interview. There was the man whose family had long since believed that tremors are followed by larger subterranean "replies"—and used the experts' assessment to convince his pregnant wife there was no need to go outside that night. All of them, including their infant son, died when the couple's home collapsed. There was the university student who was crushed to death, even though his friends had inquired about their dormitory's seismic stability just a week before. Local officials had told them not to worry.

Assembling this testimony was essential to Picuti's argument for manslaughter. He reasoned that the deaths were caused by things the scientists and engineers said and did not say. He cherry-picked past statements and snippets from scientific journals to argue that the scientists had known there would almost certainly be a major event. They couldn't have known an exact date and time, of course, says Picuti. But they deliberately suppressed discussion of risk for the sake of reassurance.

During the 13-month trial, the prospect of being found guilty struck the men as preposterous. They didn't dismiss the situation as unserious: people had died, after all. But the notion that they were guilty of negligence—of *manslaughter*—simply defied reason. "Nobody really believes that I would be so stupid as to go to Italy's most seismically active region to say to the people, 'Don't worry,'" Boschi told me.

But Picuti deftly used the scientists' own work against them. He showed the court a seismic hazard map of Italy produced by INGV, where Boschi and Selvaggi worked. Using a color-coded system from deep red (highest risk) to pale green (lowest), the map shows the probability of a major earthquake over the next 50 years.

At first glance two locations look downright dangerous: one near the southernmost part of the country, the other a wedge-shaped area that runs directly over L'Aquila. Picuti contrasted this level of risk portrayed by the map with Boschi's "improbable" assessment.

When I asked Picuti about the map, its 50-year time horizon, and the fact that day-to-day risk of a major earthquake remains low even in the areas colored deepest red, he was dismissive. "The people hear 'low' or 'improbable,' and then think: low." He told the judge that what the public should have heard is what it says on the map: highest.

But Picuti's masterstroke was wielding a 1995 academic paper called "Forecasting When Larger Crustal Earthquakes Are Likely to Occur in Italy in the Near Future." Using historical records, geological evidence, and the best seismic data available at the time, seismologists tried to predict earthquakes for different areas of Italy over time scales of 5, 20, and 100 years. According to the model, the probability that L'Aquila would get hit with a major earthquake within all of those ranges was 1. That is not a typo. The model predicted a major earthquake in L'Aquila with 100 percent certainty.

The lead author of the paper? Enzo Boschi.

To hammer home this point, Picuti put another INGV seismologist on the stand. He summarized what Boschi and his colleagues had written but then surprised Picuti by explaining that the model was simply wrong.

A formula that produced a 100 percent chance of an earthquake occurring in the next five years would obviously give the same forecast for the next 20 or 100 years, he said. Yet no major seismic event had occurred during that initial five-year window. The quake's nonarrival doesn't mean that in year six it's overdue and that in year seven it's even more so. All it means is that the model itself is wrong. Boschi and his coauthors had even flagged their conclusion as suspect in the paper itself.

As Selvaggi watched this testimony unfold, he couldn't help

feeling hopeful. Picuti had trucked out evidence that earthquake prediction isn't possible and then let a highly credentialed scientist deliver a lesson in probability and the scientific method to an audience that evidently needed one.

But Selvaggi was too optimistic. On October 22, 2012, Judge Marco Billi, an athletic 43-year-old with short-cropped black hair, walked to the front of the makeshift courtroom. Italy does not use juries: the decision was for Billi alone. Eyes down, he read his verdict in a barely audible monotone. For delivering "inexact, incomplete, and contradictory information," the scientists and engineers were found guilty of involuntary manslaughter. They each received a six-year prison sentence, pending appeal.

As Billi saw it, the attendees of that 2009 meeting were responsible for the deaths. Not of all the victims—only those who Picuti could show had a habit of fleeing their houses when there was a tremor. The science underlying Boschi's 1995 paper was of no interest to Billi. As he later told me, "We didn't look at the details of the model. We only looked at what he [Boschi] wrote—that is, that there was a probability of 1 that L'Aquila will have a major earthquake. That's all. It's Boschi's words!"

Boschi is furious over this mind-set. It's not only that the earlier model was faulty, that Picuti can't understand probabilities to save his life, or even Billi's staggering contortion of logic. It's that we shouldn't be here in the first place, talking about research and scientific papers, the whole point of which is to share so that others can disprove or refine what you've come up with. "I am willing to go to jail for this point," he thunders. "A scientist can write whatever opinions he wants in a scientific paper and it is off limits to a judge."

Even in the land of Berlusconi and the judicial circus of cases like Amanda Knox's, convicting a bunch of geoscientists in the wake of a natural disaster marks a new low. What would Galileo say? But what happened in L'Aquila is a window onto how we think about, communicate, and live with risk, and about impediments to clear thinking that afflict us all.

In the winter of 1951 a group of CIA analysts filed report NIE 29–51. Its aim: to examine whether the Soviets would invade Yugoslavia. And the bottom line? "Although it is impossible to determine which course the Kremlin is likely to adopt, we believe . . .

that an attack on Yugoslavia in 1951 should be considered a serious possibility." Once finalized, the report made its way into the bureaucratic machine.

A few days later a State Department official met up with the intelligence whiz whose team had composed the report. What did "serious possibility" mean? The CIA man, Sherman Kent, said he thought maybe there was a 65 percent chance of an invasion. But the question itself troubled him. He knew what "serious possibility" meant to him, but it clearly meant different things to different people. He decided to survey his colleagues.

The result was shocking. Some thought it meant there was an 80 percent chance of invasion; others interpreted the possibility as low as 20 percent.

Years later, Kent published an article in *Studies in Intelligence* that used the Yugoslavia report to illustrate the problem of ambiguity, particularly when talking about uncertainty. He even proposed a standardized approach to the language used for risk analysis—*probable* to indicate 75 percent confidence, give or take about 12 percent, *probably not* for 30 percent confidence, give or take about 10 percent, and so on.

Kent's risk matrix never caught on, but the need for precise "words of estimative probability" is as relevant today as it was then. After the intelligence debacle that led to the Iraq War, for example, the U.S. Office of the Director of National Intelligence reworked Kent's approach with new guidelines about the language of estimation. The goal was to minimize the blurring of information, probabilities, and confidence that had misled Americans into fretting about fictitious weapons of mass destruction.

Similarly, after Hurricane Sandy the National Oceanic and Atmospheric Administration looked at the consequences of its communications before the storm made landfall. Based on the National Weather Service's classification system, Sandy went from being a hurricane to a post-tropical cyclone. That may be good for weather nerds and lexicographers, but for the media and the public the change simply served to confuse, especially when reporters mistakenly used the word *downgraded* to explain the change.

A May 2013 audit of NOAA's performance emphasized that the agency should steer clear of abstract descriptions and instead make better use of language focusing on effects: possible flooding in these areas; storm surges up to this or that height; winds strong enough to down 100-year-old trees.

"If someone says 'unlikely' about an earthquake, you don't really know what they're talking about," says Baruch Fischhoff, professor of social and decision sciences at Carnegie Mellon. Is it unlikely-like-having-twins unlikely, or unlikely-like-winning-the-lottery unlikely?

This fuzziness especially applies when talking about potential effects from natural disasters, because events like earthquakes and category 5 hurricanes are so infrequent. What will an earthquake of magnitude 6.2 mean for your life, your street, your kids' school, your house's foundation?

Conventional wisdom tells us that people are terrible with numbers. But as Kent realized back in the 1950s, we are even worse with words. In one study that Fischhoff coauthored, people had trouble understanding a 30 percent chance of rain. It wasn't the probability that tripped them up but the word *rain.* Are we talking drizzle or downpour? All day or just part of the day? And over what area, exactly? (Communicating forecasts in Italian is extra challenging. In English we can use *forecast* instead of *prediction* to convey uncertainty. In Italian there is only *previsione,* which has a strong deterministic connotation.)

The divine cruelty of what happened in L'Aquila is that when Boschi said that a major earthquake was "improbable," he was—and remains—correct. But where a career scientist hears the word *improbable* and knows that rare events do occur, a nonscientist hears *improbable* as shorthand for *ain't gonna happen.*

Yet even the most carefully crafted communication from the Serious Risks Commission would likely have fallen short. Not because it would have failed to reach people or been met with suspicion, but because probabilities mess with our heads.

Even if we can comprehend a 30 percent chance of rain, or near-term odds like a coin flip, low-probability events are different. They have a "bewildering" effect on people, says Howard Rachlin, a professor emeritus of psychology at Stony Brook University. So we tend to lump them together; 1 in 10,000 sounds just as bewildering as 1 in 100,000. This is why people buy lottery tickets, even though the likelihood of winning is outrageously less likely than an event like a big earthquake in a seismically active region.

"All low chances seem the same," Rachlin says.

When it comes to living our lives today or making plans for next weekend, behaving as if low probability is essentially zero chance isn't necessarily a bad thing. We would be paralyzed otherwise.

But stretch that low probability over time—which is how earthquake risk is estimated—and confusion with low probabilities morphs into complete incomprehension. If you live in an earthquake-prone place for 10,000 days, the *cumulative probability* gets higher and higher, approaching 1 in 1. Our minds, unfortunately, have a hard time keeping up.

"We don't see how these small things add up when you do them over and over again," says Fischhoff. In study after study—looking at compound interest, unsafe sex, driving without a seatbelt, floods, earthquakes—we underestimate such cumulative effects. It's one of those cognitive shortcomings calling out for a name. Maybe it should be called something like *time-risk blindness.*

How alert should we be to the influence of these blind spots? When the stakes are nonexistent—in a focus group meeting about how to market saucepans, say—there is no reason to be on guard for biases that lead our thinking astray. But when the stakes are high, says Fischhoff, like when communicating seismic risk, "we owe it to people to understand what the specific barriers are and how we can best get past them."

This is where the scientists and engineers of the Serious Risks Commission went wrong, even if they didn't realize it. They had no sense of how their words would land. They were used to closed-door meetings, and the commission's mandate was to advise the Civil Protection Department, not the public. But once microphones and cameras were added into the mix, everything changed: they were now risk communicators, and whether they knew it or not, or what they might have felt about it, became irrelevant. (Unfortunately, says Fischhoff, another robust result in social science is that "people tend to exaggerate how well they communicate.")

Yet they *had* to speak up. Someone had to. If they didn't, there would have been no countermessage to the false information that was infecting the community, thanks to a one-man panic driver.

Every few mornings during the winter and spring of 2009, Giampaolo Giuliani parked his gray '96 Audi 80S on the curb of a narrow street in the tiny village of Coppito, just outside L'Aquila, and hurried into his shop. Past the cluttered piles of old keyboards, boxes of cables and cords, shelves filled with motherboard parts, and a haphazard assortment of electronics for sale, he made his way down a chipped stone staircase into his basement "lab."

His seismometer, about 20 inches tall, sat in the corner of the room, cordoned off by four old wooden chairs. On the other side of the space: a workstation, topped with a 10-year-old computer monitor, a scribble-filled notebook, and a thick instruction book titled *Costruzioni Apparecchiature Elettroniche Nucleari* (Nuclear Engineering Electronic Equipment). On the floor to the right of the workstation is a dark box. About the size of two shoeboxes, it is made of lead and, Giuliani claims, is more sensitive than similar detection equipment used by scientists the world over.

Giuliani is 67, with a serious smoker's cough and large sad eyes. Unlike soothsayers wielding crackpot ideas about crystals or signs of the zodiac, his theories are buttressed by logic. Earthquakes involve colossal amounts of energy. With such huge forces at work, it's conceivable that there would be a connection between major seismic activity and measurable changes in gases percolating up from below. Think of it as a geochemical heads-up. Giuliani has focused his attention on radon, a heavy, radioactive gas that exists in higher concentrations over geologic faults.

Seismologists have scrutinized radon for more than a generation to see if changing levels can indicate incoming quakes—and found nothing. Susan Hough, a seismologist at the Southern California Earthquake Center and the author of *Predicting the Unpredictable: The Tumultuous Science of Earthquake Prediction,* sums it up: there is "no statistically significant evidence for a relationship between radon anomalies and earthquakes."

But that didn't stop Giuliani from becoming a go-to media source on seismic activity in L'Aquila. During those tense months of the swarm, his ominous messages helped local media inject tension into their coverage of the tremors. "I gave my number to a few people, but within a month or two it felt like all of Italy had it," he says. His work as a technician at a nearby research facility for particle physics and nuclear astrophysics gave his pronouncements a veneer of institutional credibility. *Technician, physicist, scientist*—whatever. Even the *New York Times* would mistakenly refer to him as a seismologist.

On March 29, 2009, the town of Sulmona, some 30 miles from

L'Aquila, was shaken by a magnitude 3.9 tremor. Giuliani wrote on his website that this event would be followed by a big earthquake in the next day or two. With nothing else to go on, many Sulmona residents chose to evacuate.

That earthquake never came. It was good news for the public, but local officials were not pleased with Giuliani. For inciting panic, he was slapped with the equivalent of a cease-and-desist order.

He never predicted the deadly earthquake in L'Aquila, despite his continuing assertions. In reality, L'Aquila's quake whisperer did nothing more than make noise about earthquake danger in a seismically active place, at a time when the ground was frequently shaking and residents' nerves were frayed.

I recently e-mailed Hough and asked what she would say to Giuliani if she were to sit down with him for a discussion. "I would frankly have to stifle an inclination to smack him," she replied.

The scientists and officials called to the now infamous meeting in L'Aquila, she said, "did not make their statements in a vacuum, but rather responded to an individual who was making very irresponsible statements, very loudly." Without Giuliani and his divinations, there would have been no debate about the meaning of the seismic swarm, no moronic statement from Abruzzo officials that no earthquake would occur, no fumbled communications by scientists, no subsequent trial. It all traces back to the lead box in Giuliani's subterranean lair.

Today lawyers for the seven men are putting the finishing touches on their arguments for an appeal set to begin in early October. While doing so, they may want to zero in on a few odd details. The scientists who took part in the press conference after the 2009 meeting insist that they never gave any blithe message of reassurance. Yet one of the strangest, if not outright suspicious, parts of this whole saga is that audio record of that press conference has disappeared, even though video footage is available.

The lawyers may also want to present some seismology data, specifically showing how often seismic swarms are followed by nothing but calm. And they could gut Picuti's causation argument by using historical examples like this one: In 1920 a seismic swarm occurred in northwestern Tuscany, just as in L'Aquila. One afternoon the tremor was so strong (magnitude 4.1) that people decided to spend the night away from home. Nothing happened that night, though, and in the morning all the men went out to work in the fields, while women returned to their households and children went to school. That's when a magnitude 6.6 earthquake struck, killing close to 200 people, almost all of them women and children. This is the problem with the custom of fleeing during a tremor: When do you come back inside?

Even if the appeal succeeds, it's a process that could take years—years that older scientists like Boschi, who is 72, or former commission chief Franco Barberi, who is 75, may not have. Already the trial and verdict have taken a personal and professional toll, lost jobs, threatened pensions and, of course, the possibility of incarceration. At least two of the men are thinking about leaving Italy altogether.

Meanwhile the fallout from the initial guilty verdict continues. Scientists and policy wonks from Boston to Jakarta worry about the effect the case will have on experts who are asked to provide an opinion. And in Italy itself, the situation now borders on farcical. Recently the reconstituted Serious Risks Commission warned of the "significant probability" of a major earthquake. No quake followed, but there was plenty of confusion as to what "significant probability" meant. If Italians aren't totally numb to risk warnings from on high, they will be soon enough.

To keep his stress in check, Selvaggi is trying to stay healthy. He cooks a lot. He quit smoking and does his best to speak openly about the case so that his frustration doesn't fester. At home, though, he tries not to bring it up in front of the kids.

Over dinner one night I asked him about the victims' testimonies. He chose not to attend the hearings on most of those days. He thought it would be more respectful that way, and he worried that if he cried while listening to accounts of loved ones dying, his reaction might be misconstrued as an expression of remorse or guilt.

Still, he knows many of their stories, especially that of Parisse, the journalist whose teenage children were killed.

Selvaggi can't fathom how he would respond if he were to lose his children. "Maybe I would kill myself," he says. "I don't know."

What he has been through, he says, is nowhere near the torment Parisse has endured. Yet he feels a kind of connection in loss. In his office at the institute he has two disabled wall clocks. A couple of years ago he pulled out the clocks' batteries and set the hands of both to display 3:32 a.m.—the time when the earthquake struck L'Aquila, when all those people died, and when life as Selvaggi knew it came to a halt.

"I have spent my life trying to understand earthquakes to help prevent harm to people," he says. "Now those people are against me, when I think we should be together."

BARRY YEOMAN

From Billions to None

FROM *Audubon*

> Men still live who, in their youth, remember pigeons; trees still live who, in their youth, were shaken by a living wind. But a few decades hence only the oldest oaks will remember, and at long last only the hills will know.
>
> —Aldo Leopold, "On a Monument to the Pigeon," 1947

IN MAY 1850 a 20-year-old Potawatomi tribal leader named Simon Pokagon was camping at the headwaters of Michigan's Manistee River during trapping season when a far-off gurgling sound startled him. It seemed as if "an army of horses laden with sleigh bells was advancing through the deep forests towards me," he later wrote. "As I listened more intently, I concluded that instead of the tramping of horses it was distant thunder; and yet the morning was clear, calm, and beautiful." The mysterious sound came "nearer and nearer," until Pokagon deduced its source: "While I gazed in wonder and astonishment, I beheld moving toward me in an unbroken front millions of pigeons, the first I had seen that season."

These were passenger pigeons, *Ectopistes migratorius,* at the time the most abundant bird in North America and possibly the world. Throughout the 19th century, witnesses had described similar sightings of pigeon migrations: how they took hours to pass over a single spot, darkening the firmament and rendering normal conversation inaudible. Pokagon remembered how sometimes a traveling flock, arriving at a deep valley, would "pour its living mass" hundreds of feet into a downward plunge. "I have stood by the grandest waterfall of America," he wrote, "yet never have my astonishment, wonder, and admiration been so stirred as when I have witnessed these birds drop from their course like meteors from heaven."

Pokagon recorded these memories in 1895, more than four decades after his Manistee River observation. By then he was in the final years of his life. Passenger pigeons too were in their final years. In 1871 their great communal nesting sites had covered 850 square miles of Wisconsin's sandy oak barrens—136 million breeding adults, naturalist A. W. Schorger later estimated. After that the population plummeted until, by the mid-1890s, wild flock sizes numbered in the dozens rather than the hundreds of millions (or even billions). Then they disappeared altogether, except for three captive breeding flocks spread across the Midwest. About September 1, 1914, the last known passenger pigeon, a female named Martha, died at the Cincinnati Zoo. She was roughly 29 years old, with a palsy that made her tremble. Not once in her life had she laid a fertile egg.

This year marks the 100th anniversary of the passenger pigeon's extinction. In the intervening years researchers have agreed that the bird was hunted out of existence, victimized by the fallacy that no amount of exploitation could endanger a creature so abundant. Between now and the end of the year, bird groups and museums will commemorate the centenary in a series of conferences, lectures, and exhibits. Most prominent among them is Project Passenger Pigeon, a wide-ranging effort by a group of scientists, artists, museum curators, and other bird lovers. While their focus is on public education, an unrelated organization called Revive & Restore is attempting something far more ambitious and controversial: using genetics to bring the bird back.

Project Passenger Pigeon's leaders hope that by sharing the pigeon's story, they can impress upon adults and children alike our critical role in environmental conservation. "It's surprising to me how many educated people I talk to who are completely unaware that the passenger pigeon even existed," says ecologist David Blockstein, senior scientist at the National Council for Science and the Environment. "Using the centenary is a way to contemplate questions like 'How was it possible that this extinction happened?' and 'What does it say about contemporary issues like climate change?'"

*

They were evolutionary geniuses. Traveling in fast, gargantuan flocks throughout the eastern and midwestern United States and Canada—the males slate-blue with copper undersides and hints of purple, the females more muted—passenger pigeons would search out bumper crops of acorns and beechnuts. These they would devour, using their sheer numbers to ward off enemies, a strategy known as "predator satiation." They would also outcompete other nut lovers—not only wild animals but also domestic pigs that had been set loose by farmers to forage.

In forest and city alike, an arriving flock was a spectacle—"a feathered tempest," in the words of conservationist Aldo Leopold. One 1855 account from Columbus, Ohio, described a "growing cloud" that blotted out the sun as it advanced toward the city. "Children screamed and ran for home," it said. "Women gathered their long skirts and hurried for the shelter of stores. Horses bolted. A few people mumbled frightened words about the approach of the millennium, and several dropped on their knees and prayed." When the flock had passed over, two hours later, "the town looked ghostly in the now-bright sunlight that illuminated a world plated with pigeon ejecta."

Nesting birds took over whole forests, forming what John James Audubon in 1831 called "solid masses as large as hogsheads." Observers reported trees crammed with dozens of nests apiece, collectively weighing so much that branches would snap off and trunks would topple. In 1871 some hunters coming upon the morning exodus of adult males were so overwhelmed by the sound and spectacle that some of them dropped their guns. "Imagine a thousand threshing machines running under full headway, accompanied by as many steamboats groaning off steam, with an equal quota of R.R. trains passing through covered bridges—imagine these massed into a single flock, and you possibly have a faint conception of the terrific roar," the *Commonwealth*, a newspaper in Fond du Lac, Wisconsin, reported of that encounter.

The birds weren't just noisy. They were tasty too, and their arrival guaranteed an abundance of free protein. "You think about this especially with the spring flocks," says Blockstein, the ecologist. "The people on the frontiers have survived the winter. They've been eating whatever food they've been able to preserve from the year before. Then, all of a sudden, here's all this fresh meat flying by you. It must have been a time for great rejoicing: the pigeons are here!" (Not everyone shouted with joy. The birds also devoured crops, frustrating farmers and prompting Baron de Lahontan, a French soldier who explored North America during the 17th century, to write that "the Bishop has been forc'd to excommunicate 'em oftner than once, upon the account of the Damage they do to the Product of the Earth.")

The flocks were so thick that hunting was easy—even waving a pole at the low-flying birds would kill some. Still, harvesting for subsistence didn't threaten the species' survival. But after the Civil War came two technological developments that set in motion the pigeon's extinction: the national expansions of the telegraph and the railroad. They enabled a commercial pigeon industry to blossom, fueled by professional sportsmen who could learn quickly about new nestings and follow the flocks around the continent. "Hardly a train arrives that does not bring hunters or trappers," reported Wisconsin's *Kilbourn City Mirror* in 1871. "Hotels are full, coopers are busy making barrels, and men, women, and children are active in packing the birds or filling the barrels. They are shipped to all places on the railroad, and to Milwaukee, Chicago, St. Louis, Cincinnati, Philadelphia, New York, and Boston."

The professionals and amateurs together outflocked their quarry with brute force. They shot the pigeons and trapped them with nets, torched their roosts, and asphyxiated them with burning sulfur. They attacked the birds with rakes, pitchforks, and potatoes. They poisoned them with whiskey-soaked corn. Learning of some of these methods, Potawatomi leader Pokagon despaired. "These outlaws to all moral sense would touch a lighted match to the bark of the tree at the base, when with a flash—more like an explosion—the blast would reach every limb of the tree," he wrote of an 1880 massacre, describing how the scorched adults would flee and the squabs would "burst open upon hitting the ground." Witnessing this, Pokagon wondered what type of divine punishment might be "awaiting our white neighbors who have so wantonly butchered and driven from our forests these wild pigeons, the most beautiful flowers of the animal creation of North America."

Ultimately the pigeons' survival strategy—flying in huge predator-proof flocks—proved their undoing. "If you're unfortunate enough to be a species that concentrates in time and space, you make yourself very, very vulnerable," says Stanley Temple, a professor emeritus of conservation at the University of Wisconsin.

Passenger pigeons might have even survived the commercial slaughter if hunters weren't also disrupting their nesting grounds —killing some adults, driving away others, and harvesting the squabs. "It was the double whammy," says Temple. "It was the demographic nightmare of overkill and impaired reproduction. If you're killing a species far faster than they can reproduce, the end is a mathematical certainty." The last known hunting victim was "Buttons," a female, which was shot in Pike County, Ohio, in 1900 and mounted by the sheriff's wife (who used two buttons in lieu of glass eyes). Almost seven decades later a man named Press Clay Southworth took responsibility for shooting Buttons, not knowing her species, when he was a boy.

Even as the pigeons' numbers crashed, "there was virtually no effort to save them," says Joel Greenberg, a research associate with Chicago's Peggy Notebaert Nature Museum and the Field Museum. "People just slaughtered them more intensely. They killed them until the very end."

Contemporary environmentalism arrived too late to prevent the passenger pigeon's demise. But the two phenomena share a historical connection. "The extinction was part of the motivation for the birth of modern twentieth-century conservation," says Temple. In 1900, even before Martha's death in the Cincinnati Zoo, Republican congressman John F. Lacey of Iowa introduced the nation's first wildlife protection law, which banned the interstate shipping of unlawfully killed game. "The wild pigeon, formerly in flocks of millions, has entirely disappeared from the face of the earth," Lacey said on the House floor. "We have given an awful exhibition of slaughter and destruction, which may serve as a warning to all mankind. Let us now give an example of wise conservation of what remains of the gifts of nature." That year Congress passed the Lacey Act, followed by the tougher Weeks-McLean Act in 1913 and, five years later, the Migratory Bird Treaty Act, which protected not just birds but also their eggs, nests, and feathers.

The passenger pigeon story continued to resonate throughout the century. In the 1960s populations of the dickcissel, a sparrow-like neotropical migrant, began crashing, and some ornithologists predicted its extinction by 2000. It took decades to uncover the reason: During winters the entire world population of the grasslands bird converged into fewer than a dozen huge flocks, which settled into the *llanos* of Venezuela. There rice farmers who considered the dickcissels a pest illegally crop-dusted their roosts with pesticides. "They were literally capable, in a matter of minutes, of wiping out double-digit percentages of the world's population," says Temple, who studied the bird. "The accounts are very reminiscent of the passenger pigeon." As conservationists negotiated with rice growers during the 1990s—using research that showed the dickcissel was not an economic threat—they also invoked the passenger pigeon extinction to rally their colleagues in North America and Europe. The efforts paid off: the bird's population has stabilized, albeit at a lower level.

Today the pigeon inspires artists and scientists alike. Sculptor Todd McGrain, creative director of the Lost Bird Project, has crafted enormous bronze memorials of five extinct birds; his passenger pigeon sits at the Grange Insurance Audubon Center in Columbus, Ohio. The Lost Bird Project has also designed an origami pigeon and says thousands have been folded—a symbolic recreation of the historic flocks.

The most controversial effort inspired by the extinction is a plan to bring the passenger pigeon back to life. In 2012 Long Now Foundation president Stewart Brand (a futurist best known for creating the *Whole Earth Catalog)* and genetics entrepreneur Ryan Phelan cofounded Revive & Restore, a project that plans to use the tools of molecular biology to resurrect extinct animals. The project's "flagship" species is the passenger pigeon, which Brand learned about from his mother when he was growing up in Illinois. Revive & Restore hopes to start with the band-tailed pigeon, a close relative, and "change its genome into the closest thing to the genetic code of the passenger pigeon that we can make," says research consultant Ben Novak. The resulting creature will not have descended from the original species. "[But] if I give it to a team of scientists who have no idea that it was bioengineered, and I say, 'Classify this,' if it looks and behaves like a passenger pigeon, the natural historians arc going to say, 'This is *Ectopistes migratorius.*' And if the genome plops right next to all the other passenger pigeon genomes you've sequenced from history, then a geneticist will have to say, 'This is a passenger pigeon. It's not a band-tailed pigeon.'"

Revive & Restore plans to breed the birds in captivity before returning them to the wild in the 2030s. Novak says the initial research indicates that North American forests could support a reintroduced population. He hopes animals brought back from extinction—not just birds but eventually also big creatures like woolly mammoths—will draw the public to zoos in droves, generating revenues that can be used to protect wildlife. "De-extinction [can] get the public interested in conservation in a way that the last forty years of doom and gloom has beaten out of them," he says.

Other experts aren't so sanguine. They question whether the hybrid animal could really be called a passenger pigeon. They doubt the birds could survive without the enormous flocks of the 19th century. And they question Novak's belief that the forests could safely absorb the reintroduction. "The ecosystem has moved on," says Temple. "If you put the organism back in, it could be disruptive to a new dynamic equilibrium. It's not altogether clear that putting one of these extinct species from the distant past back into an ecosystem today would be much more than introducing an exotic species. It would have repercussions that we're probably not fully capable of predicting."

Blockstein says he wanted to use the 100th anniversary as a "teachable moment." Which eventually led him to Greenberg, the Chicago researcher, who had been thinking independently about 2014's potential. The two men reached out to others until more than 150 institutions were on board for a year-long commemoration: museums, universities, conservation groups (including Audubon state offices and local chapters), libraries, arts organizations, government agencies, and nature and history centers.

Project Passenger Pigeon has since evolved to be a multimedia circus of sorts. Greenberg has published *A Feathered River Across the Sky,* a book-length account of the pigeon's glory days and demise. Filmmaker David Mrazek plans to release a documentary called *From Billions to None.* At least four conferences will address the pigeon's extinction, as will several exhibits. "We're trying to take advantage of every possible mechanism to put the story in front of audiences that may not necessarily be birdwatchers, may not necessarily even be conservationists," says Temple.

The commemoration goes beyond honoring one species. Telling the pigeon's story can serve as a jumping-off point for exploring the many ways humans influence, and often jeopardize, their own environment. Today an estimated 13 percent of birds are threatened, according to the International Union for Conservation of Nature. So are 25 percent of mammals and 41 percent of amphibians, in large part because of human activity. Hydropower and road construction imperil China's giant pandas. The northern bald ibis, once abundant in the Middle East, has been driven almost to extinction by hunting, habitat loss, and the difficulties of doing conservation work in war-torn Syria. Hunting and the destruction of wetlands for agriculture drove the population of North America's tallest bird, the whooping crane, into the teens before stringent protections along the birds' migratory route and wintering grounds helped the wild flock build back to a few hundred. Little brown bats are dying off in the United States and Canada from a fungus that might have been imported from Europe by travelers. Of some 300 species of freshwater mussels in North America, fully 70 percent are extinct, imperiled, or vulnerable, thanks to the impacts of water pollution from logging, dams, farm runoff, and shoreline development. Rising sea temperatures have disrupted the symbiotic relationship between corals and plant-like zooxanthellae, leading to a deadly phenomenon called coral bleaching. One third of the world's reef-building coral species are now threatened.

If public disinterest helped exterminate the passenger pigeon, then one modern-day parallel might be public skepticism about climate change. In an October poll by the Pew Research Center for the People and the Press, only 44 percent of Americans agreed there was solid evidence that the earth is warming because of human activity, as scientists now overwhelmingly believe. Twenty-six percent didn't think there was significant proof of global warming at all. In another Pew poll, conducted last spring, 40 percent of Americans considered climate change a major national threat, compared with 65 percent of Latin Americans and slimmer majorities in Europe, Africa, and the Asia-Pacific region.

This denial of both the threat and our own responsibility sounds eerily familiar to those who study 19th-century attitudes toward wildlife. "Certainly if you read some of the writings of the time," says Blockstein, "there were very few people who put stock in the idea that humanity could have any impact on the passenger pigeons." (Audubon himself dismissed those who believed that "such dreadful havoc" as hunting would "soon put an end to the species.") Today attitudes toward climate change sound similar, continues Blockstein. "It's the same kind of argument: 'The world is so big and the atmosphere is so big; how could we possibly have an impact on the global climate?'"

Even the political rhetoric of those who don't want to address climate change aggressively has 19th-century echoes. "The industry that paid people to kill these birds said, 'If you restrict the killing, people will lose their jobs,'" notes Greenberg—"the very same things you hear today."

Project Passenger Pigeon might not change the minds of hard-core climate skeptics. For the rest of us, though, it could serve as a call to take responsibility for how our personal and collective actions affect wildlife and climate. Maybe a close look at the history of human folly will keep us from repeating it.

Contributors' Notes

Other Notable Science and Nature Writing of 2014

Contributors' Notes

Jake Abrahamson is a writer living in California.

Burkhard Bilger has been a staff writer at *The New Yorker* since 2000. He was a senior editor at *Discover* from 1999 to 2005, and from 1994 to 1999 a deputy editor and writer at *The Sciences,* where his work helped earn two National Magazine Awards and six nominations. His book, *Noodling for Flatheads* (2000), was a finalist for the PEN/Martha Albrand Award. Bilger is a Branford Fellow at Yale University, from which he graduated in 1986. He is at work on a book about his grandfather's experiences during the Second World War.

Sheila Webster Boneham writes about animals, environment, gender, and culture in the anthropological sense. She is as interested in the questions we ask as in the answers. Her publications include seventeen nonfiction books, four novels, and a number of essays, short stories, and poems. Her books have won the Maxwell Award for Fiction and for Nonfiction and the MUSE Award for Nonfiction. The essay included here won the Prime Number Magazine Award for Creative Nonfiction and has been nominated for the Pushcart Prize.

Rebecca Boyle is an award-winning journalist who grew up in Colorado, a mile closer to space. A former political reporter, she now writes about the vast realm of science, from astronomy to zoonoses. She focuses on bats and spiders for her blog, Eek Squad, which is hosted by *Popular Science.* Her work appears in *Wired, New Scientist, Ask, Aeon,* and many other publications for adults and kids. Boyle lives in St. Louis with her husband and daughter.

Alison Hawthorne Deming's most recent book is *Zoologies: On Animals and the Human Spirit* (2014). She is the author of three additional nonfiction books and four poetry books, most recently *Rope* (2009), with *Stairway to Heaven* due out in 2016. She is Agnese Nelms Haury Chair in Environment and Social Justice and professor in the Creative Writing Program at the University of Arizona and a 2015 Guggenheim Fellow.

Sheri Fink is the author of the *New York Times* best-selling book *Five Days at Memorial: Life and Death in a Storm-Ravaged Hospital* (2013), winner of the National Book Critics Circle Award for nonfiction, the Ridenhour Book Prize, the J. Anthony Lukas Book Prize, the Los Angeles Times Book Prize, the Southern Independent Booksellers Alliance Book Award, the American Medical Writers Association Medical Book Award, and the NASW Science in Society Journalism Book Award. Fink's news reporting has been awarded the Pulitzer Prize, the National Magazine Award, and the Overseas Press Club Lowell Thomas Award, among other journalism prizes. A former relief worker in disaster and conflict zones, Fink received her MD and PhD from Stanford University. Her first book, *War Hospital: A True Story of Surgery and Survival* (2003), is about medical professionals under siege during the genocide in Srebrenica, Bosnia-Herzegovina. She is a correspondent at the *New York Times.*

Atul Gawande practices general and endocrine surgery at Brigham and Women's Hospital and is a professor at the Harvard Chan School of Public Health and Harvard Medical School. He is also executive director of Ariadne Labs, a joint center for health systems innovation, and chairman of Lifebox, a nonprofit organization making surgery safer globally. He has been a staff writer for *The New Yorker* since 1998. He is the author of four *New York Times* bestsellers: *Complications, Better, The Checklist Manifesto,* and, most recently, *Being Mortal: Medicine and What Matters in the End.*

Writer and photographer **Lisa M. Hamilton** focuses on agriculture and rural communities. She is the author of *Deeply Rooted: Unconventional Farmers in the Age of Agribusiness* and has written for *Harper's Magazine, McSweeney's, Virginia Quarterly Review,* and *The Atlantic.* Her current work is about crop genetic resources in the era of climate change.

Rowan Jacobsen writes for *Harper's Magazine, Outside, Mother Jones, Orion,* and other magazines. His *Outside* piece "Heart of Dark Chocolate" received the Lowell Thomas Award from the Society of American Travel Writers for adventure story of the year; his *Eating Well* piece "Or Not to Bee" received a James Beard Award; his *Harper's* piece "The Homeless

Herd" was named magazine piece of the year by the Overseas Press Club; and his *Outside* piece "Spill Seekers" was selected for *The Best American Science and Nature Writing 2011.* He was a 2012 Alicia Patterson Foundation Fellow, writing about endangered cultures on the borderlands between India, Myanmar, and China, and is a 2015 McGraw Fellow, writing about the promise of fake meat. He is also the author of six books, including *Fruitless Fall, The Living Shore,* and *Shadows on the Gulf.*

Leslie Jamison has written a novel, *The Gin Closet,* and a collection of essays, *The Empathy Exams.* Her work has appeared in *Harper's Magazine, Oxford American, A Public Space, Virginia Quarterly Review, The Believer,* and the *New York Times,* where she is a columnist for the *Sunday Book Review.* She lives in Brooklyn.

Brooke Jarvis is an independent journalist who focuses on longform narrative and environmental reporting. She is the author of "When We Are Called to Part," published by *The Atavist,* and has written features for a long list of magazines. Her work has been supported by the Middlebury Fellowship in Environmental Journalism and the Alicia Patterson Foundation. She lives in Seattle.

Sam Kean is the *New York Times* best-selling author of *The Tale of the Dueling Neurosurgeons, The Disappearing Spoon,* and *The Violinist's Thumb.* He has given well over a hundred talks on his books, in five different countries and in two dozen different states.

Jourdan Imani Keith has been awarded fellowships from Wildbranch, Santa Fe Science Writing Workshop, VONA, Hedgebrook, and Jack Straw. She received funding from Artist Trust, 4Culture, and Seattle's Office of Arts and Culture for her choreopoem/play, *The Uterine Files,* and *Coyote Autumn,* a memoir. Seattle Poet Populist emerita and Seattle Public Library's first naturalist in residence, she is a storyteller and student of Sonia Sanchez. An excerpt from her memoir is included in the travel writing anthology *Something to Declare.* She is at work on a series of linked essays called *Tugging at the Web,* an expansion of her TEDx Talk.

Eli Kintisch is a contributing correspondent for *Science* magazine. A two-time MIT Knight Science Journalism Fellow, he covers climate change, oceans, and the Arctic and has written for *Slate, Nautilus, New Scientist,* and the *Los Angeles Times.* His 2010 book, *Hack the Planet: Science's Best Hope—or Worst Nightmare—for Averting Climate Catastrophe,* received a starred review from *Publishers Weekly.*

Elizabeth Kolbert is a staff writer for *The New Yorker* and the author of *The Sixth Extinction,* which won the 2015 Pulitzer Prize. Her series on global warming, "The Climate of Man," won the American Association for the Advancement of Science's magazine writing award and a National Academies communications award. Those articles became the basis for *Field Notes from a Catastrophe: Man, Nature, and Climate Change.* She is a two-time National Magazine Award winner and has received a Heinz Award and a Lannan Literary Fellowship. Kolbert lives in Williamstown, Massachusetts.

Amy Maxmen writes for *Newsweek, Nature,* the *New York Times, Nautilus,* and many other publications. She's interested in the entanglements of evolution, medicine, policy, and the people behind the research. With fellowships from the Pulitzer Center on Crisis Reporting, she has reported several stories from Africa. Prior to writing she earned a PhD from Harvard University in evolutionary and organismic biology.

Seth Mnookin is an associate professor of comparative media studies/writing and the associate director of the Graduate Program in Science Writing at MIT. His most recent book, *The Panic Virus: The True Story of the Vaccine-Autism Controversy,* won the National Association of Science Writers 2012 Science in Society book award. He is also the author of the 2006 national bestseller *Feeding the Monster,* about the Boston Red Sox, and 2004's *Hard News,* about the *New York Times.* He lives with his wife, their two children, and their ten-year-old adopted pit bull in Boston, Massachusetts, and can be found online at sethmnookin.com and @sethmnookin.

Dennis Overbye is a science reporter for the *New York Times.* He is the author of *Einstein in Love: A Scientific Romance,* which was a Los Angeles Times Book Prize finalist, and *Lonely Hearts of the Cosmos: The Story of the Scientific Quest for the Secret of the Universe,* a finalist for the National Book Critics Circle Award and the Los Angeles Times Book Prize, and the winner of the American Institute of Physics award for science writing. "Chasing the Higgs," an article he wrote for the *New York Times,* was a finalist for the 2014 Pulitzer Prize. He lives in Manhattan with his wife, Nancy Wartik, and their daughter.

Matthew Power died in March 2014 while on an assignment in Uganda for *Men's Journal.* He was thirty-nine. Described by his editor at *Outside* magazine as "relentlessly generous," he regularly encouraged young writers. His articles have been anthologized in *The Best American Travel Writing, The Best American Spiritual Writing,* and now, to our good fortune, here.

Sarah Schweitzer is a feature writer for the *Boston Globe.* She joined the paper's staff in 2001 after working for the *St. Petersburg Times* and the *Concord Monitor.* She was a Pulitzer Prize finalist in feature writing in 2015 for "Chasing Bayla." She lives in New Hampshire with her husband, two children, and two dogs.

Michael Specter has been a staff writer at *The New Yorker* since 1998. He writes often about science, technology, and global public health. Since joining the magazine, he has written several articles about the global AIDS epidemic, as well as about avian influenza, malaria, the world's diminishing freshwater resources, synthetic biology, the attempt to create edible meat in a lab, and the debate over the meaning of our carbon footprint. He has also published many "Profiles," of subjects including Lance Armstrong, the ethicist Peter Singer, Sean (P. Diddy) Combs, Manolo Blahnik, and Miuccia Prada. Specter went to *The New Yorker* from the *New York Times,* where he had been a roving foreign correspondent based in Rome. From 1995 to 1998 Specter served as the *Times'* Moscow bureau chief. He went to the *Times* from the *Washington Post,* where from 1985 to 1991 he covered local news before becoming the *Post*'s national science reporter and, later, the newspaper's New York bureau chief.

Specter has twice received the Global Health Council's annual Excellence in Media Award, first for his 2001 article about AIDS, "India's Plague," and second for his 2004 article "The Devastation," about the ethics of testing HIV vaccines in Africa. He received the 2002 AAAS Science Journalism Award for his 2001 article "Rethinking the Brain," on the scientific basis of how we learn. His most recent book, *Denialism: How Irrational Thinking Hinders Scientific Progress, Harms the Planet, and Threatens Our Lives,* received the 2009 Robert P. Balles Annual Prize in Critical Thinking, presented by the Committee for Skeptical Inquiry. In 2011, Specter won the World Health Organization's Stop TB Partnership Annual Award for Excellence in Reporting for his *New Yorker* article "A Deadly Misdiagnosis," about the dangers of inaccurate TB tests in India, which has the highest rate of TB in the world. Specter splits his time between Brooklyn and upstate New York.

Meera Subramanian is an award-winning journalist whose work has been published in the *New York Times, Nature, Virginia Quarterly Review,* and elsewhere. She is also the author of *A River Runs Again: India's Natural World in Crisis, from the Barren Cliffs of Rajasthan to the Farmlands of Karnataka* (2015) and an editor for the online literary magazine *Killing the Buddha.* She lives on Cape Cod in Massachusetts. Find her at www.meerasub.org and @meeratweets.

Kim Todd is the author of *Sparrow; Chrysalis: Maria Sibylla Merian and the Secrets of Metamorphosis;* and *Tinkering with Eden: a Natural History of Exotic Species in America.* Her work has received the PEN/Jerard Award and the Sigurd Olson Nature Writing Award, and she teaches literary nonfiction in the MFA program at the University of Minnesota. "Curious" was written with the help of a residency at the Mesa Refuge.

David Wolman is an award-winning journalist, author, and speaker. He is a contributing editor at *Wired* and the author of four works of nonfiction: *Firsthand, The End of Money, Righting the Mother Tongue,* and *A Left-Hand Turn Around the World.* He has also written for such publications as the *New York Times,* the *Wall Street Journal, The New Yorker, Time, Nature,* and *Outside.* He lives in Oregon with his wife and two children. Follow him @davidwolman.

Barry Yeoman is a freelance journalist who specializes in putting a human face on complex issues, including science and the environment. The winner of numerous national awards, he was named by *Columbia Journalism Review* as one of nine investigative reporters who are "out of the spotlight but on the mark." His writing appears in *Audubon, onEarth, Saturday Evening Post,* and *The American Prospect* and can be read at barryeoman.com. He lives in Durham, North Carolina.

Other Notable Science and Nature Writing of 2014

SELECTED BY TIM FOLGER

SANDRA ALLEN
How Climate Change Will End Wine as We Know It. *BuzzFeed,* November 20.

BRETT ANDERSON
Louisiana Loses Its Boot. *Medium,* September 8.

NANCY C. ANDREASEN
Secrets of the Creative Brain. *The Atlantic,* July/August.

NATALIE ANGIER
Too Much of a Good Thing. *The New York Times,* September 22.

PETER B. BACH
The Day I Started Lying to Ruth. *The New York Times,* May 6.

DAVID P. BARASH
God, Darwin, and My Biology Class. *The New York Times,* September 27.

MARCIA BARTUSIAK
Dark Matters. *Natural History,* January.

RICK BASS
The Constant Garden. *Audubon,* November/December.

KENNETH BROWER
Quicksilver. *National Geographic,* March.

CHIP BROWN
Kayapo Courage. *National Geographic,* January.

KENNETH CHANG
A Lost and Found Nomad Helps Solve the Mystery of a Swimming Dinosaur.
The New York Times, September 11.

Emily Eakin
The Excrement Experiment. *The New Yorker,* December 1.
Gregg Easterbrook
What Happens When We All Live to 100? *The Atlantic,* October.
Mieke Eerkins
Seep. *Creative Nonfiction,* Spring.

Douglas Fox
The Dust Detectives. *High Country News,* December 22.
Edward Frenkel
The Lives of Alexander Grothendieck, a Mathematical Visionary. *The New York Times,* November 25.

Ted Genoways
Hog Wild. *On Earth,* Spring.
Dana Goodyear
Death Dust. *The New Yorker,* January 20.
Denise Grady
An Ebola Doctor's Return from Death. *The New York Times,* December 7.
Bruce Grierson
What if Age Is Nothing but a Mind-Set? *The New York Times,* October 22.
Liza Gross
Face of Hope. *Discover,* September.
Elizabeth Grossman
Banned in Europe, Safe in the U.S. *Ensia,* June 9.

Lisa M. Hamilton
The Quinoa Quarrel. *Harper's Magazine,* May.
Alan Heathcock
Zero Percent Water. *Medium,* September 21.
Robin Marantz Henig
Guess I'll Go Eat Worms. *Medium,* July 21.

George Johnson
Why Everyone Seems to Have Cancer. *The New York Times,* January 4.
James L. Johnson
Frank Malina: America's Forgotten Rocketeer. *IEEE Spectrum,* July 31.

Jonathon Keats
Playing the Meteorite Market. *Discover,* June.
Raffi Khatchadourian
A Star in a Bottle. *The New Yorker,* March 3.
Elizabeth Kolbert
Eclipse. *Orion,* November/December.
Robert Kunzig
Carnivore's Dilemma. *National Geographic,* November.

Tom Levitt
What's the Beef? *Earth Island Journal,* Winter.
Sharon Levy
The Bees' Needs. *On Earth,* Summer.
Jane Braxton Little
The Wolf Called OR7. *Appalachia,* Winter/Spring.

Phil McKenna
New Life for the Artificial Leaf. *Ensia,* November 17.
Rebecca Mead
Starman. *The New Yorker,* February 17 & 24.
Greg Miller
Inside the Strange New World of DIY Brain Stimulation. *Wired,* May.
Peter Miller
Digging Utah's Dinosaurs. *National Geographic,* May.

Nick Neely
Slow Flame. *Harvard Review Online,* April 22.
Michelle Nijhuis
Caviar's Last Stand. *Food and Environment Reporting Network,* July 28.

Jennifer Oullette
Multiverse Collisions May Dot the Sky. *Quanta Magazine,* November 10.
Stephen Ornes
Archimedes in the Fence. *The Last Word on Nothing,* May 23.
Dennis Overbye
A Picture Captures Planets Waiting to Be Born. *The New York Times,* December 18.

Ross Perlin
Endangered Speakers. *n+1,* Spring.
Tom Philpott
California Goes Nuts. *Mother Jones,* November/December.
Stephen Porder
World Changers. *Natural History,* November.
Lewis Pugh
Swimming Through Garbage. *The New York Times,* September 28.
Robert Michael Pyle
Free-Range Kids. *Orion,* November/December.

Shruti Ravindran
What Solitary Confinement Does to the Brain. *Aeon,* February 27.
Julie Rehmeyer
The Immortal Devil. *Discover,* May.
Jim Robbins
Building an Ark for the Anthropocene. *The New York Times,* September 27.

SHARMAN APT RUSSELL
Meet the Beetles. *Orion,* November/December.

NICHOLAS ST. FLEUR
The Future Looks Bleak for Bones. *The Atlantic,* December 23.
MEGAN SCUDELLARI
Never Say Die. *Medium,* May 7.
JESSICA SEIGEL
America's Getting the Science of Sun Exposure Wrong. *Nautilus,* Fall.
CHRIS SOLOMON
Rethinking the Wild. *The New York Times,* July 6.
MICHAEL SPECTER
Against the Grain. *The New Yorker,* November 3.
Partial Recall. *The New Yorker,* May 19.
DAWN STOVER
Living on a Carbon Budget: Or You Can't Always Get What You Want. *Bulletin of the Atomic Scientists,* October 5.
EDWARD STRUZIK
The End and Beginning of the Arctic. *Ensia,* December 2.

DEB OLIN UNFERTH
Cage Wars. *Harper's Magazine,* November.

ERIK VANCE
Gods of Blood and Stone. *Scientific American,* July.
On Call in the Wild. *Discover,* March.
PAUL VOOSEN
Striving for a Climate Change. *The Chronicle of Higher Education,* November 3.
Wasteland. *National Geographic,* December.
We Are All Mutants. *The Chronicle of Higher Education,* March 24.

SCOTT WEIDENSAUL
Have Lemmings, Will Travel. *Audubon,* March/April.
E. O. WILSON
On Free Will. *Harper's Magazine,* September.
NATALIE WOLCHOVER
In a Grain, a Glimpse of the Cosmos. *Quanta Magazine,* June 13.
A New Physics Theory of Life. *Quanta Magazine,* January 6.